WHY
GEOGRAPHY
MATTERS

By Nicholas Crane

Clear Waters Rising: A Mountain Walk Across Europe

Two Degrees West: An English Journey

Mercator: The Man Who Mapped the Planet

Great British Journeys

Coast: Our Island Story

The Making of the British Landscape: From the Ice Age to the Present

Nicholas Crane is an author, geographer, cartographic expert and recipient of the Royal Scottish Geographical Society's Mungo Park Medal in recognition of outstanding contributions to geographical knowledge, and of the Royal Geographical Society's Ness Award for popularising geography and the understanding of Britain. He has presented several acclaimed series on BBC2, among them *Map Man*, *Great British Journeys*, *Britannia*, *Town* and *Coast*. He served as President of the Royal Geographical Society from to 2015 until 2018.

WHY GEOGRAPHY MATTERS

*A Brief Guide to
the Planet*

NICHOLAS CRANE

WEIDENFELD & NICOLSON

First published in Great Britain in 2018 by Weidenfeld & Nicolson
This paperback edition published in 2020 by Weidenfeld & Nicolson
an imprint of The Orion Publishing Group Ltd
Carmelite House, 50 Victoria Embankment,
London, EC4Y 0DZ

An Hachette UK Company

1 3 5 7 9 10 8 6 4 2

A CIP catalogue record for this book is
available from the British Library.

ISBN (Mass Market Paperback) 978 1 4746 0830 5
ISBN (ebook) 978 1 4746 0831 2

Typeset by Input Data Services Ltd, Somerset

Printed and bound in Great Britain by Clays Ltd, Elcograf S.p.A.

www.orionbooks.co.uk
www.weidenfeldandnicolson.co.uk

Dedicated to all students and teachers of geography

CONTENTS

PREFACE

It doesn't seem long ago that I was a boy on a bike with a map in the countryside. A child of the Cold War pedalling the land of the East Angles in search of overgrown Roman towns and abandoned American airfields. Much has changed since then. Venta Icenorum has been cleared of thorns, the 389th Bomb Group is a car factory and my village of coal-burning, draughty homes has been swallowed by a city with clean air and cycle lanes. The pace of change increases. The pedals spin faster. That's the nature of geography. Places change. Environments change. People change.

This book was first published in October 2018. Since then, the world has become a different place. So this is an updated edition with an updated title.

Why Geography Matters celebrates a life-enhancing subject and argues that geography urgently needs to be reinstalled as humanity's operating system.

Geography explains how the world works. It explores places and the interactions between people and the environment. In schools and universities, it's often separated into two dynamic fields: the cultures, societies and economies of human geography, and the landscapes and environment of physical geography.

Geography works through time as well as space. It connects hunter-gatherers with cyberculture. Long before the internet, we knew so much about geography that we evolved into the most successful species on the planet. Without a basic understanding of geography, we couldn't feed ourselves, find our way about, or enjoy the wonders of existence on this spinning orb 94 million miles from the Sun. It is a beautiful subject.

Why Geography Matters is a plea for geography to be rebooted in our collective imaginations. All of the planet's most pressing challenges – from rising oceans to decimated wildlife, from droughts to floods, from inequality to poverty, from security to sustainability – are essentially geographical. They are all related to the environment, to places and to people. We need to know the basics of geography in the same way that we know how to speak, read and write. For the indefinite future, geographical knowledge will guide us along our life-paths.

The story of human existence on this spinning rock-ball has been one of accelerating change. In a finite space occupied by a swelling population, interactions multiply. And we are beginning to understand where the limits lie. The destruction of natural landscapes for meat and dairy farming, the burning of fossil fuels for energy, the pollution of rivers and oceans, soils and air, the ceaseless cruelties of war and poverty, all have their tipping points.

In the couple of years since I wrote the first edition of this book, we have crept closer to the edge. In the UK, the 'State of Nature 2019' report revealed that 600 species have dropped in population by 13 per cent since 1970. The same year, a summary of insect studies concluded that declines in almost all regions of the world may lead to the extinction of 40 per cent of insects over the coming decades. Insects matter because they support many of our key ecosystems. Fish populations are falling. Plastics pollute seas and rivers. Despite the signing in 2014 of a UN declaration aimed at halving deforestation by 2020, the rate of tree cover loss has gone up by 43 per cent. Around the world, 64 million acres of forest – that's about the size of the UK – are being lost every year. One-third of the world's topsoils are degraded. The war in Yemen has brought 460,000 cases of cholera in the first half of 2019 alone. As I write this, 22 million Yemenis need humanitarian aid and protection. Children learn of this world in geography lessons. And they see it.

Children also know why the climate is changing faster than it was one hundred years ago. The forest fires in Siberia, California and the Amazon are symptoms of a heating planet. And so are rising sea levels, more destructive monsoons in India, more violent hurricanes in the Atlantic and typhoons in the Pacific. The system is stirring with more force. The Arctic continues to warm at twice the global rate, shrinking the sea ice and reducing the size of the planet's reflective shield. The thawing permafrost releases ever more methane, a greenhouse gas 34 times more potent that CO_2. In Antarctica, the amount of glacier mass lost between 2015 and 2019 was the highest for any five-year period on record.

'Every week brings us news of more climate-related devastation,' observed the UN Secretary General in 2019. Human activities have caused 1 °C of global warming since pre-industrial times. At the current rate of warming, we're likely to hit 1.5 °C by around 2040. After that, things get a lot more expensive. A 2018 report from the UN's Intergovernmental Panel on Climate Change advised that limiting warming to 1.5 °C is possible within the laws of chemistry and physics, but that it 'would require unprecedented transitions in all aspects of society.'

'Unprecedented transitions' have become a hot topic. Institutions ranging from the International Monetary

Fund to the World Economic Forum and the Bank of England are calling for reform of a global economic system that has collided with our planetary system. At the 2019 UN Climate Action Summit, the Governor of the Bank of England, Mark Carney, warned that firms who fail to align their business models with the transition to net-zero greenhouse gases would 'cease to exist'. Months earlier, Sir David Attenborough warned of 'irreversible damage to the natural world and the collapse of our societies.' Public polls show a doubling of citizens concerned about the environment. Fear of the future has become a mobilising emotion. New movements like Extinction Rebellion warn of 'extreme ecological collapse' and demand 'revolutionary changes'. The young Swedish climate activist Greta Thunberg has led millions of the world's school students onto the streets of 150 countries. 'I want you to panic,' she told the World Economic Forum at Davos in January 2019. 'I want you to feel the fear I feel every day. I want you to act.'

She is right. We have to act. The surest way to avert fear and panic is by taking prompt, appropriate action. Seen through a geographical orb, everything is connected. We are part of a dynamic system that links air, land, water and life. Simultaneously, we must reduce our impact on natural systems and prepare for change. As never before, we all need to know how the world works. We need to be familiar with the basic geography that describes our places and the interactions between people and the environment. We need to take informed action as citizens, voters, policy-formers and leaders. That is the message of this book.

Why Geography Matters explores geography through six themes. In the first chapter, I pull back from the partial pictures familiar to us as terrestrial bipeds and think about our planet and its complex systems as a whole. The second chapter looks at the way water shapes our physical world, from the Amazon to Antarctica, from the Maeander to the Ganges, from river to ocean to atmosphere. In the third chapter, I take a trip through our human world, from Mumbai to Beijing, from the inner migrant within our heads to multimillion megacities. After that, I go back in time to explore the origins of our geographical intuition and – by way of Africa, the Arctic and China – our evolution as spatial specialists. In the fifth chapter, I track mapping revolutionaries from the Euphrates to the Nile to the Golden State. Chapter six is being written by us all. The disruption of Earth's natural systems by *Homo sapiens* demands a deeper, wider understanding of geography. Our planet is mid-chapter and we are the authors of its outcome.

Nicholas Crane
London, 2019

ONE

The View from L1

L et's begin by setting the scene.

From a million miles away our planet is a blue, cloud-whorled sphere set against a matt black void. This cute selfie is updated every few hours by a spacecraft at the L1 Lagrange Point, a cosmic sweet-spot where the combined gravitational forces of the Sun and its only habitable planet equal the centrifugal force exerted on a satellite. It is known as a 'neutral gravity point' and is one of five Lagrange Points that exist in the Sun–Earth force fields. Of these points of equilibrium, L1 is unique because it is the one location that allows satellites to observe both the Sun and a fully illuminated Earth. The 2-metre craft that arrived at L1 in 2015 after a voyage lasting 110 days was NASA's first operational satellite in deep space.

Among the instruments on board the Deep Space Climate Observatory (DSCOVR) is a four-megapixel

1

camera. Several times a day, the Earth Polychromatic Imaging Camera (EPIC) takes a set of images in ten different wavelengths, which can be combined to create colours that make sense to the human brain. The camera's remote-controlled iris captures pictures that outshine the original 'blue-marble' portraits snapped on a hand-held Hasselblad by the crew of Apollo 17 as they flew to the Moon in 1972. Viewing EPIC's high-resolution images in the White House, President Obama tweeted that NASA had composed a 'beautiful reminder that we need to protect the only planet we have'. Here was Earth at its most improbable: a bright orb of life suspended in the infinite, celestial night. For all practical purposes, Earth is a one-off.

Our planet's journey from lifeless fireball to organic marvel is a story of constant change. We began with a bang as a star exploded and hurled a cloud of hot dust and gas into the Orion Spur of the Milky Way. Out of this galactic inferno whirled the beginnings of a solar system, a yellow dwarf star at the centre of a spinning disc of debris that congregated into asteroids and comets, planets and moons. When we look at the night sky, we see the progeny of gravity. Early Earth was a hot ball that cooled over 50 million years or so into an inner core and mantle of molten material and an outer, cooler crust. Wrapping the planet was

Earth from L1, with the Moon

a thin film of gases that included hydrogen sulphide, methane and carbon dioxide. From 3.8 billion years ago, water was collecting in cooled depressions on the planet's crust.* By 3.5 billion years ago, rocks were recording death in the form of stromatolites, fossils formed from sticky mats of bacteria that had accumulated in shallow waters where sunlight allowed them to photosynthesise. By 2 billion years ago, oxygen was among the gases enveloping the planet and new types of microfossils were forming in the crust. Algae and soft-bodied animals followed. The land was colonised. Fishes swam in the seas. Amphibians took to land. Insects and plants appeared, then reptiles.

On at least five occasions, life was hammered on a planetary scale. Of these mass extinctions, the most devastating occurred 251 million years ago. The geological traces of this catastrophe have been tracked from Siberia to Canada and Greenland and from Australia to South East Asia. Most revealing of all have been a set of sedimentary beds in South China, at Meishan, where a layer of limestone loaded with fossils is sealed with ash and clay. The Meishan rock sequence has been chosen as the 'world type section' (the locality used as the universal standard for a geological

---

* With apologies to readers who prefer more specific timescales, I'm providing approximations.

boundary) for the base of the Triassic, the geological period that followed the collapse of the Permian. Given its causes, there is understandable interest in this particular extinction. It all happened very quickly. In their *Science* paper of 2011, Shu-zhong Shen and his colleagues identified layers of charcoal and soot at several locations in South China.

Beset by sudden aridity, the world's rainforests had dried and flared into flame. Unprotected soils were subject to destructive erosion and virulent fungus. Oceans became anoxic. The Permian mass extinction wiped out 96 per cent of the world's species. The likely cause was a series of volcanic eruptions in Siberia that triggered a sudden release of carbon dioxide and methane, causing a runaway greenhouse effect and extreme global warming. It was the biggest species crash in Earth's history and it took 20 million years for ecosystems to recover.

More recently, a mass extinction was inflicted by a meteorite in southern Mexico. Around 65 million years ago, a lump of space rock 10 kilometres wide collided with the Yucatán Peninsula, excavating a 150-kilometre crater and flinging a vast column of dust into the upper atmosphere. Sunlight was blocked, causing temperatures to plummet. Plants couldn't photosynthesise. Large reptiles died of starvation or cold. The dinosaurs went, along with huge, fish-like ichthyosaurs, the

Diversity in the Cretaceous

17-metre marine mosasaurs, predatory plesiosaurs and winged pterosaurs. Power was no signifier of survival. The atmospheric catastrophe at the end of the Cretaceous period rebooted evolution's operating system and opened ecological voids that were occupied by the animal-ancestors of *Homo sapiens*.

Fast-forward to East Africa around 3 million years ago and we find ourselves in the spectacular landscapes of the Great Rift Valley. Two of Earth's tectonic plates are moving apart, creating a massive rent in the crust which has been occupied by a mosaic of dense tropical forest and open grassland. Lurking in the shadows of cliff and bough is a smart ape that has adapted to both savannah and forest, that can climb a tree and use its 'legs' to balance upright while reaching for high-hanging fruit. It can also scamper upright across open ground. *Australopithecus afarensis* also seems to have been making and using tools.

The evolution of versatile proto-humans such as *A. afarensis* was encouraged in part by climate variability and environmental disruption. Around 2.6 million years ago, there was a switch to glaciation in the northern hemisphere and to drier, more arid conditions in the tropics. The fall in temperature marked the onset of a thermal rollercoaster: the Pleistocene epoch. Out at L1, a viewer would have noted reflective white shields waxing and waning at Earth's axial extremes.

It has happened around twenty-two times during the Pleistocene.

While alternate episodes of extreme glaciation and relatively balmy interglacials became the cyclic pattern in the northern hemisphere, conditions in East Africa grew more intense. Over a million or so years, forest gave way to savannah grasslands, while the East African climate swung between intense extremes of hyper-aridity and overabundant rainfall. This new world of climate variability increased the rate of extinctions and speciation, the winners being those who could handle environmental disruption. Into this changing Africa strode the genus *Homo*.

Some species of *Homo* were more successful than others. Tool-wielding *Homo erectus* – whose brain may have been as large as 1,100 cubic centimetres and who stood up to 1.6 metres tall – broke all records for survival of a human species; 'Upright Man' walked out of Africa and endured for nearly 2 million years. Early examples have been found in Georgia, southern Europe and eastern China. *Homo soloensis* was adapted to surviving in the tropics. *Homo neanderthalensis* was adapted for cold climates and spread across much of Europe using tools and fire, but was overtaken by *Homo sapiens*, who had a smaller, 1,400-cubic-centimetre, brain and arrived between 300,000 and 200,000 years ago.

The most recent thermal crash and recovery reveals how adaptable humans have had to be. Only 22,000 years ago, Canada and northern Europe were compressed beneath 4 kilometres of blue ice. The heights of East Asia and the southern Andes were clamped by ice caps. At the peak of the last glacial period 20,000 years ago, temperatures on land had fallen by 20 °C, and so much of the planet's water was locked up as ice that global sea levels had fallen by more than 100 metres. Islands – and even continents – became conjoined by land bridges. People could walk between Asia and America; Britain was connected to Europe.

Right now, we're in an interglacial period that kicked in around 11,700 years ago, when an episode of extreme climate change catapulted temperatures upward and removed the great ice sheets from North America and Europe. For a while, there were lakes of fresh water and lush vegetation in the Sahara Desert. Subsequent setbacks have failed to stem the rise of *Homo sapiens*, although catastrophic events, from volcanic eruptions to hurricanes, floods, landslides, disease and wars, have exacted a terrible toll. Increasingly, climate change is being linked to headline disasters. 'Paradoxically,' writes Jamie Woodward of Manchester University, 'in an era of warming climate, the study of the ice age past is now more important than ever.'

So, it's taken some 4.6 billion years for a cosmic swirl

of gas and dust to evolve into a planet fit for butterflies and children. The blue dot in the black cosmos now hosts an estimated 8.7 million species.

This intricate web of life exists because Earth operates as a system in which everything is connected and interdependent. The word 'complex' does not begin to describe the workings of this system. Indeed, it is so diverse and complicated that no computer currently comes close to creating a model. Such is the urgency of the challenge that thousands of scientists are working on what the geologist Richard Alley dubbed in 2000 an 'Operator's Manual' for the planet: a clearer understanding of how the various components of the Earth's system work, and 'how they are wired together and depend on each other'.

What is it about this sphere of interacting systems circling the Sun that makes it possible for a kingfisher to flash across the River Ant, or for a yak to forage the grassland of Tibet? One way of simplifying the Earth's system is to think of it as four connected components, or 'spheres': the lithosphere, atmosphere, hydrosphere and biosphere – land, air, water and life.

The lithosphere – the 'sphere of rock' – is the uppermost layer of Earth and comprises both the crust itself and the upper layer of the mantle. It is separated into a number of tectonic plates that move at a rate of a few centimetres a year. There was a time when Africa was connected to South America, North America and Antarctica. Collisions between the plates have created mountain ranges like the Himalayas and Alps. But we know remarkably little about the ground beneath our feet. Despite our ability to send a spacecraft on a 7.8-billion-kilometre trip to Saturn, the deepest artificial hole we have created in the Earth's crust is only 12 kilometres, drilled by Soviet geologists on the Kola Peninsula and abandoned in 1992. The heat and pressure were so extreme that the hole squeezed closed as the drill was withdrawn. On the surface of the lithosphere, a range of climates have contributed to wonderfully varied landscapes. Of the Earth's land area, one-fifth is defined as desert and receives less than 25 centimetres of rain per year. The hot deserts that girdle central latitudes are decorated with astonishing landforms: sand dunes shaped like crescents, whose curving crests grow feathers in the wind; mushroom rocks undercut by flying sand grains; stippled plains of gravel; cliffs chiselled with dry chasms.

The second component of Earth's system is the layer of gases held by gravity to Earth's surface – the atmosphere. Compared to the diameter of the planet, it is thin: above 3,000 or so metres in the mountains I begin to notice the lack of oxygen, and by 6,000 metres I'm panting like a steam engine. The base of the atmosphere, the troposphere, reaches up to 10,000 metres, and it is within this layer that most of our weather events, clouds and precipitation occur. Above the troposphere is the stratosphere, which climbs to 50 kilometres, and above that is the mesosphere, which goes up to 85 kilometres. Near the top of this layer, temperatures fall to −90 °C. Beyond the mesosphere is the thermosphere, home of the International Space Station, orbiting 330 kilometres from Earth. And at the outer limits of the atmosphere is the exosphere, which is so airless that some scientists consider it part of space. Mixed within the atmosphere are gases essential for life, among them nitrogen (78 per cent by volume), oxygen (21 per cent), water vapour (1 per cent) and carbon dioxide (0.04 per cent).

The third component in the system is the hydrosphere: water in the form of liquid, vapour and ice. This includes all oceans and seas, together with fresh-water bodies such as lakes, rivers and streams, moisture found in soil and rocks, atmospheric moisture, ice crystals, permafrost, sea ice, ice sheets and glaciers. About 10 per cent of the planet is covered by glacial ice. Around 71 per cent of Earth's surface is covered by water, but virtually all of it − 97 per cent − is seawater. About 80 per cent of the ocean's volume is below 5 °C, but surface waters in the tropics rise to as much as 30 °C. The salinity of seawater has been fairly constant for a few hundred million years, but way back in geological time there was less salt in the seas.

The fourth and final 'sphere', the biosphere, is the one that contains all living organisms, from fungi, plants and animals to people like you and me. Virtually all of this life exists in a thin, busy band not much higher than 200 metres above ground and 3 metres beneath it. There are different ways of separating the living planet into 'biomes', geographical areas that can be defined by their species. At its simplest, the world has two biomes: one water-based and the other on land. But biomes can also be classified as marine, fresh-water, grassland, forest, desert and tundra. Still further subdivisions can be created by separating, for example, forest into deciduous, tropical and taiga. The World Wildlife Fund uses a classification of fourteen major terrestrial biomes. Oddly, perhaps, for a species that has discovered how to fit a computer into a watch, we don't know for sure how many species our biomes contain. The figure of 8.7 million I used above comes from a scientific paper published in 2011, which also concluded that 91 per cent of species in the oceans and 86 per cent on land still await description. Just under one-third of Earth's land area is clothed with forest, burgeoning green belts teeming with life; just over one-third is agricultural land. The amount of land occupied by towns and cities is difficult to estimate because there is no widely agreed definition of 'urban land'. But in 2014, four academics achieved a minor breakthrough by devising three measures of urbanisation. Zhifeng Liu and his colleagues found that 'global urban land' (urban area delineated by administrative boundaries) amounted to 3 per cent of the Earth's surface.

Now, here's the complication. This quartet of 'spheres' exists in a state of constant, mutual interaction. The surface of the lithosphere, for example, is continually being modified by erosion, weathering and the shifting of material by the rains, rivers and tides of the hydrosphere. At various times in the past, volcanic spluttering of the lithosphere has vented ash into the atmosphere, altering the climate – and thus the hydrosphere – and disrupting the biosphere.

The atmosphere functions as a natural thermostat, responding to energy flows that govern temperatures on Earth. About one-third of the incoming short-wave radiation from the Sun is reflected into space, while the rest is absorbed by the ocean and the land, which radiate the energy as long-wave radiation. Some of this is absorbed by greenhouse gases like water vapour, carbon dioxide, methane and nitrous oxide, which has the effect of warming the atmosphere. Without the greenhouse gases, Earth would be too cold for human beings.

The natural cycle of the hydrosphere – the water

cycle – sends water on a round trip through the other three 'spheres': from the ocean into the atmosphere as water vapour, which condenses as rain or snow to fall on the mountains and plains of the lithosphere, where it nourishes the biosphere en route to the ocean. The geological carbon cycle moves carbon between oceans, land and atmosphere, where chemical reactions create carbon stores for extended periods. Rocks, oceans and plants are all carbon stores (also known as stocks, pools and reservoirs). There is a balance in the cycle between carbon production and absorption, but this can be interrupted by events such as volcanic eruptions, which sometimes spew vast amounts of carbon into the atmosphere.

In the biosphere, every living thing, from corals to polar bears, from farmers to bankers, from ants to antelopes; and every biome, from mangrove to desert, exists in a state of perpetual interaction with the other 'spheres'.

The science that embraces the planet and its spheres is geography, a word inherited from the Greek for 'earth description'. Over the millennia, it's been a subject that has fascinated *Homo sapiens*, a field of curiosity that helps us make sense of our world, its people, its places, its environments.

TWO

Water World

In this chapter I'd like to explore the wonders and complexities of one of the four spheres. I've chosen the hydrosphere.

A long time ago, I was lucky enough to take a short voyage in a dugout canoe down the headwaters of the Amazon. Most nights, fierce spears of rain turned the river into a mottled, hissing snake as raindrops began an odyssey to their ocean terminus. But there was, of course, no terminus. The moment a raindrop joins a river is not a beginning, but one of countless waymarks on a never-ending cycle. Once in the ocean, that drop is evaporated and rises as water vapour to form clouds, which condense into falling raindrops. It's a closed system. The total amount of water in the world does not change. Every raindrop belongs to the most beautiful recycling system on Earth.

Let's join some water in an upland stream and

follow it through a complete cycle. Only 1.6 per cent of the world's surface and atmospheric fresh water is moved by rivers. Far more – 67.4 per cent – is stored in lakes. Volatile and beautiful, the physical form of a river, from bounding stream to wandering meander, mimics our own passage through life. Geographers speak of 'young', 'mature' and 'old' rivers. In *Landmarks*, Robert Macfarlane's book of 'place-words', he explores the rich English, Welsh and Gaelic lexicon for moving water, from *drindle*, for a minor trickle, to *pistyll* for a waterfall and *burraghlas* for a 'torrent of brutal rage'. And it's in these youthful stages that rivers express their vitality most vigorously: the V-shaped valley clutching its ribbon of crystal light; the cataracts and pools at boundaries in the bedrock; the waterfalls that mark a tough old sill overlying layers of softer stone; the gorge where waterfalls have retreated for millennia. Underground, streams trickle and seep and thunder through the crust's hidden plumbing, feeding subterranean aquifers and water tables, dissolving and abrading passages and caverns into 3D cave systems that branch and squirm for kilometres.

The river has always been a creator of human geography. It is a conduit for human interaction, a linear attractant for hunters and gatherers, for camping and for cooking hazelnuts, for erecting staked roundhouses and processional ways, for building civilisations and launching an Industrial Revolution:

> It is from the midst of this putrid sewer that the
> greatest river of human industry springs up and
> carries fertility to the whole world. From this foul
> drain pure gold flows forth. Here it is that humanity
> achieves for itself both perfection and brutalization,
> that civilization produces its wonders, and that civ-
> ilized man becomes again almost a savage.

In Alexis de Tocqueville's description of Manchester in
1835, the polluted River Irwell becomes a metaphor for
the city's 'capricious creative force'.

In our contemporary age of unspent time, rivers
succour the human spirit. They always have. (It isn't
hard to imagine *Homo neanderthalensis* wondering
about the meaning of life while gazing into a trout
pool.) They're places for swimming and fishing and
floating, reliable muses for painters, writers and poets.
When Imtiaz Dharker portrays London's river 'trailing
marshdamp and the warmth / of creatures it has slept
with all these years', she's summoning the ghost of a
river past, a river I've known all my life and one that
tells me a thousand stories, from Mesolithic mudbank
burials to Saxon boatmen and vanquished bankers.

Words gather as silt, the folios thickest downstream.
On the coast of Turkey there was once a necklace of

Greek colonies and a city of thinkers called Miletus, and beside it the River Maeander. Like the Euphrates, the Tigris and the Nile before it, the Maeander developed its own story, a mythological presence in the narratives of its own people and of poets distant in space and time. Of the river to Miletus, Quintus of Smyrna wrote:

> *Maeander's flood deep rolling swept thereby,*
> *Which from the Phrygian upland, pastured o'er*
> *By myriad flocks, around a thousand forelands*
> *Curls, swirls, and drives his hurrying ripples on*
> *Down to the vine-clad land of Carian men.*

For generations, the river flowed through Greek and Roman imaginations. Strabo claimed that it was the Maeander, 'its course so exceedingly winding', that gave rise to the adjectival use of 'meandering'. Ovid wrote of the Maeander 'playing along its winding path', and Pliny claimed that the river was 'so tortuous that it is often believed to turn and flow backwards'. Eventually, so much silt came down the Maeander that the river choked to death. Today, the ruins of Miletus can be found more than 10 kilometres from the coast.

The portals through which rivers pass at the coast are libraries of geographical stories. The word 'delta' comes down to us as the Greek letter Δ, whose shape resembles the spreading fan of the Nile. As river water enters the sea, the coarser part of its load is dropped immediately, but the finer particles are carried further, out along the various distributary channels of the delta. If the amount of sediment deposited by the river exceeds the amount removed by coastal processes, the delta will grow into a range of shapes that are dependent upon variables such as the supply of material, wave action and tidal currents. Where the river's natural levees extend along the sides of distributary streams, a characteristic 'bird's-foot delta' develops. And so it is with the Mississippi, 'stuck out over the Gulf of Mexico like a fishing rod', as Mark Twain put it. By contrast, the Nile is an arcuate – arc-shaped – delta, and so is the Ganges, the world's largest delta and home to over 100 million people.

And so the river reaches the ocean, with its rims of water-formed architecture: structures built by deposition, from spits to beaches, bars and forelands, to those hewn through erosion, from cliffs to arches, stacks, stumps and caves.

I spent ten years presenting more than eighty BBC films about coasts and it made me – a London landlubber – revisit my notions of 'edge'. With time, I came to see the physical limits of land as a threshold. The more I learned of our eternal attachments to the sea, the less constrained I felt by land, an emotion familiar to the Tongan writer Epeli Hau'ofa, who reminded his readers that the indigenous people of Oceania saw the world differently: 'Their universe comprised not only land surfaces, but the surrounding ocean as far as they could traverse and exploit it.' It was 'continental men, Europeans and Americans, who drew imaginary lines across the sea, making colonial boundaries that, for the first time, confined ocean peoples to tiny spaces'. The Pacific ceased to be 'a sea of islands' and became 'islands in a far sea . . . tiny, isolated dots in a vast ocean'.

Everyone who has taken to the planet's blue expanses in a small boat knows the sea moves in accordance with complex forces of wind and tide and current. On the grandest scale is the global stream known as the thermohaline circulation or global conveyor belt, a tangled, branching, looping system of currents that distribute energy from solar heating across all the major oceans, transport atmospheric carbon dioxide to the deep and bring nutrients to the surface, where they support local ecosystems and fisheries.

We end our journey on a languid afternoon aboard a Brixham sailing trawler en route to France – a random image provided by my memory bank of voyages. Evaporating from the warm ocean beyond the trawler's scarred hull is water vapour, gathering as fleets of clouds that include some of the molecules that had mottled an Amazonian river twenty years earlier. And so the hydrological cycle continues.

There is another, vital part of the hydrosphere. Interacting with the fluid water systems are the ice systems: the cryosphere. The world's largest ice sheet is Antarctica, a 14-million-square-kilometre sub-zero mirror storing 60 per cent of Earth's fresh water. Antarctica is the freak freezer, a polar continent. The thickness of Antarctica's ice varies because it rests on an uneven surface. The eastern Antarctic ice sheet is draped over mountains and valleys, but parts of the western sheet are more than 2,500 metres below sea level.

Images of Antarctica invariably convey the impression of frozen immobility, but this frigid world is riven with slow-grinding glaciers and quicker 'ice streams', where rates of flow reach 1,000 metres a year. Some of these ice streams are 50 kilometres wide, 2,000 metres thick and flow for hundreds of kilometres, often lubricated at their base by water. Where ice sheets meet the sea, they float as ice shelves because ice is less dense than water. When parts break off, they form icebergs. So Antarctica operates as an open system, with inputs and outputs from and to other systems. Fallen snow accumulates as ice in glaciers and eventually melts into the ocean.

Glaciation is one of the planet's most spectacular agents of beauty. How could anyone gaze upon Yosemite or the Matterhorn or Cerro Torre or Machapuchare and not gasp at the creative interaction of natural forces? At the spectacle of icefalls cascading between lacerated peaks? One of many ways in which ice sculpts these glaciated landscapes is through frost shattering: when water freezes, it expands by around 9 per cent, prising apart cracks in rock. A still, cold night in the mountains can be rent by detonations as blocks split, fall and shatter. Sharp, angular fragments of rock turn the undersides of glaciers into gigantic rasps that gouge the U-shaped valleys of school textbooks. Because the Pleistocene Ice Age is so recent, at least in geological terms, and because glaciation is an ongoing process, many elements of the landscape bear the fresh brush strokes of cryospheric artistry: the 'corrie' (or cwm or cirque), for example, a deep hollow on a mountainside with a cliff-like back wall and a raised lip that can act as a dam for the still waters of a small lake; the pyramidal peak, where the back walls of three or more corries meet in a vertiginous spike; the arête, where adjacent glaciers have cut into a mountainside, creating a tilted, knife-edged ridge; hanging valleys, where a glacier has sliced off the lower end of a valley leaving a suspended Shangri La. And then there is the catalogue of smaller features: glacially smoothed hummocks; polished slabs striated with scratches indicating the direction of glacial travel; strips of loose stone like high-tide marks along the sides of valleys; dams pushed into place by glacier snouts and massive boulders that have been picked up by long-gone glaciers and dumped far away, where they sit to this day like incongruous Pleistocene artworks. The 20-ton lump of andesite that rests in monumental splendour within the main quadrangle of Manchester University began its journey 80 miles away in the mountains of the Lake District. Glacial 'erratics' are story-makers from other ages and places. In the midst of the soft, level coastal sediments of the Netherlands, a former island in the Zuiderzee is embellished with huge blocks of stone carried by ice from Norway. Off Highway 7, about an hour out of Calgary, the Okotoks erratic rests on the prairie like a smashed megalithic warehouse. It's big – 16,500 tons – has its own parking lot and is thought to have piggy-backed a 300-kilometre glacier ride from Jasper National Park.

This, of course, is not the whole story. I haven't, for example, mentioned the role in the water cycle of percolation or vegetation storage, or many other crucial components of this closed system. But we need to move on. The water cycle is changing. Earlier in this interglacial, even after we'd switched to agriculture, the water cycle was relatively intact, a virtuous circle of aquatic transfer in which waters ran clear and salmon swam far inland, a system more or less in balance. But that is no longer true.

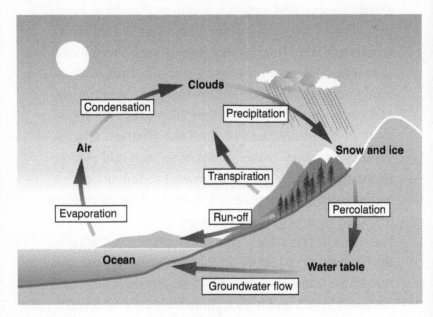

The global hydrological cycle

*

Human dependence on fire burst into an inferno with the onset of the Industrial Revolution. Fuel that had been banked under ground for hundreds of millions of years went up in smoke. As coal, oil and natural gas were burned, their stored carbon was released into the atmosphere as carbon dioxide and methane, both potent greenhouse gases. The effect has been to raise average global temperatures, which are triggering changes in the Earth system.

One of the crucial roles of polar ice is to reflect incoming solar energy back to space. The reflectivity of a surface is known as its 'albedo'. Ice, snow and clouds have a high albedo of 40 to 80 per cent and help prevent Earth from overheating. Tarmac, for example, has a low albedo of 5 to 10 per cent. Since the end of the Ice Age, both poles have been surrounded by frozen sea. In the past, Antarctica would double in size each winter as around 20 million square kilometres of sea froze; in the Antarctic summer, this salty ice would reduce to around 3 million square kilometres. But sea ice around Antarctica is becoming erratic. In March 2017, summer sea ice shrank to 2 million square kilometres, its lowest extent since satellite observations began in 1979.

In the Arctic, the cycle of freeze and thaw also appears to be undergoing major disruption. In an epic of cross-correlation, a team of American scientists used a variety of historical sources to track at a monthly time resolution the extent of Arctic sea ice back to 1850, and they found no precedent on the pan-Arctic scale for the reductions seen in the twenty-first century. This is not news to Greenlanders. Reminiscing in 2017, the author and former President of the Inuit Circumpolar Council, Aqqaluk Lynge, said: 'We used to go out on the ice by dog sled around December but for the last fifteen years there has been no ice at all.'

Summer sea ice in the Arctic has been reducing at a rate of 10 per cent per decade since 1979. Currently, it's unclear when the Arctic Ocean will reach an 'ice-free' state, often defined as a coverage of less than a million square kilometres. James Overland and Muyin Wang of the USA's National Oceanic and Atmospheric Administration (NOAA) have opted for 'the first half of the twenty-first century, with a possibility of major loss within a decade or two'. The 2013 Fifth Assessment Report of the Intergovernmental Panel on Climate Change (IPCC) took a consensus view that Arctic ice would linger till later this century. The British Antarctic Survey has said that 'it might be possible' to avert an ice-free Arctic Ocean if the temperature goals of the Paris Agreement are met. In *A Farewell to Ice* (2016), the sea-ice scientist Peter Wadhams criticised the models used by the IPCC and – using the University of Washington's Pan-Arctic Ice Ocean Modeling and Assimilation System (PIOMAS) – predicted that the Arctic Ocean's 'death spiral' would terminate at a 'drop-dead date of about 2020'. Wadhams thinks that once the Arctic Ocean suffers its first ice-free month, the open-water season will extend to four or five months 'within a few years'. The two most troubling impacts of Arctic sea-ice loss are the reduction of albedo, which Wadhams estimates will have a global warming effect equivalent to the last twenty-five years of carbon-dioxide emissions, and the thawing of the Arctic seabed, which will release plumes of methane, whose greenhouse effect is twenty-three times greater per molecule than that of carbon dioxide.

Antarctica's glaciers are also shedding ice. Way back in 1968, John Mercer, a glaciologist at Ohio State University, warned his peers that melting ice caps had the capacity to cause rapid sea-level rise. A decade later, Mercer reiterated his warning, writing that 'deglaciation may be part of the price that must be paid in order to buy enough time for industrial civilisation to make the changeover from fossil fuels to other sources of energy'.

Today, warming oceans and atmosphere are thinning glaciers, increasing ice loss. Some of Antarctica's largest glaciers have accelerated by as much as 50 per cent and are draining so quickly into the ocean that they are significant contributors to global sea-level rise. Five adjacent glacier catchments that collectively drain one-third of the West Antarctic ice sheet have doubled their rate of ice loss in only six years and are now contributing around 10 per cent of global sea-level rise. Along the edge of the Antarctic Peninsula, a series of ice shelves have been breaking up: in 1995 the enormous Larsen A ice shelf collapsed, followed in 2002 by the Larsen B shelf. In 2014, a gigantic rift opened in the Larsen C shelf, and in July 2017 an iceberg the size of Delaware broke loose. Overall, the amount of fresh water locked up in Antarctic ice is enough to raise global sea level by 70 metres.

At the other end of the world, a combination of surface melting and iceberg calving is rapidly diminishing the overall ice mass of Greenland. Between 2002 and 2016, the annual ice loss was around 280 gigatons, enough to raise global sea levels by 0.8 millimetres a year. Evidence is accumulating, too, that Greenland's melting ice is perhaps the reason why the northern extension of the thermohaline circulation is weakening. While a dying Gulf Stream will not trigger the apocalyptic scenarios depicted in the climate-disaster movie *The Day After Tomorrow*, studies suggest that climates could be affected and deep-ocean ecosystems transformed. More research is urgently required.

Accelerated sea-level rise has humanitarian consequences. The increase in average global temperatures has swollen the volume of the world's oceans through thermal expansion and the melting of polar ice sheets and mountain glaciers. Between 1870 and 2010 sea levels rose by 21 centimetres, and now that the system's energy has shifted, waters will continue to rise. Climate scientists estimate that average sea levels will increase by between 30 centimetres and 1 metre by 2100. The inhabitants of Kiribati, a necklace of thirty-three low-lying islands in the Pacific, are preparing to lose their homeland over the next fifty years. In the first climate relocation of its kind, President Anote Kong of Kiribati has negotiated the purchase of a block of land 2,000 kilometres away across the Pacific on one of the Fijian islands. Fiji itself is confronted by a bill of $4.5 billion – an amount equivalent to its entire GDP – over the next decade if it is to keep pace with the effects of rising seas. The most affected parts of the globe will be densely populated, low-lying coastal areas. In the Ganges–Brahmaputra mega-delta, sea-level rise could affect as many as 3 million people by 2050. A worst-case scenario would see Bangladesh lose nearly one-quarter of its land area by the end of the century. And Europeans were given a wake-up call in 2017 during the UN Climate Conference in Bonn, which identified particular risks for Greece, Belgium and the Netherlands. Venice is spending €5.5billion on installing fifty-seven flood barriers.

Besides sea-level rise, a warmer Earth will increase the frequency of heatwaves, droughts and floods. In August 2003, an area of high pressure hovered over western Europe for much of the month. As temperatures rose, the River Danube fell to its lowest level in 100 years, revealing unexploded bombs and tanks from the Second World War. Rivers and reservoirs feeding public water supplies and hydro schemes ceased to flow or ran low. As foliage lost its water content, forest fires broke out across the continent. In Portugal, an area equivalent to Luxembourg burned. In the Alps, accelerated melting caused rock falls. Heat-related deaths across Europe rose to 70,000. France, the country worst hit, experienced a 60 per cent increase in mortality, despite the mobilisation of a *plan chaleur extrème*, an extreme heat plan. It is anticipated that the summer temperatures of 2003 could become 'normal' by 2050.

Droughts are expected to occur more frequently. In the past they have lasted for months, or even years. From 2011 till 2012, East Africa suffered the most severe drought in sixty years. Crops failed, livestock was decimated and cereal prices leaped. Some 13.3 million people needed water, food and emergency health care and hundreds of thousands were driven from Somalia by drought and conflict. A research paper published in 2013 by three scientists from the UK Met Office

Hadley Centre concluded that 'human influence' was found to increase the probability that rainy seasons would be 'as dry as, or drier than, 2011'. One of the heightened risks associated with climate change is that regions vulnerable to drought will experience it more frequently, while regions unfamiliar with drought may be struck by it. The extreme drought of 2015–17 came close to causing the city of Cape Town to run out of water and it pushed some 50,000 people below the poverty line.

Floods are also predicted to increase. The ones that struck Pakistan in 2010 were triggered by the most damaging monsoon rains in the country's history. The Indus burst its banks and flood waters sped south through Punjab, Balochistan and Sindh, inundating one-fifth of the country's land area and killing 1,985 people, affecting 18 million and damaging or destroying 10,000 schools. The following year, monsoon flooding destroyed over a million houses and affected 6 million. The water was back for a third year running, when floods in 2012 cost at least 450 people their lives and affected over 4.8 million. And the waters returned in 2013. The monsoon rains of August 2017 killed 1,200 across India, Bangladesh and Pakistan, and affected 40 million. Streets in Mumbai became waist-deep rivers. Climate simulations suggest that future monsoons are going to be more severe as a warming Indian Ocean allows more moisture to be carried north to India across a steepening thermal contrast between land and sea.

Climate is not the water cycle's only difficulty. Deforestation has left a hydrological imprint ever since the tree-felling habits of hunter-gatherers unbonded topsoil, which was carried into rivers where it left drifts of silt. Today, deforestation is exerting a far greater impact on the water cycle. At the local level, widespread felling accelerates the run-off of rainwater and causes erosion, landslides and downriver flooding. Communities exposed to the greatest risk of flooding are often relatively poor, with a scarcity of data, but research undertaken in Indonesian Borneo found that deforestation had increased the frequency of floods. Over three years, more than 750,000 people had been displaced by flooding. The areas most affected were found to be those subject to deforestation for mining and oil-palm plantations.

Deforestation has the capacity to affect the water cycle in other ways, too. In the floodplain lakes of the Amazon basin, fish stocks are being reduced because felling has removed 'tree food': leaves, fruits, insects and miscellaneous detritus eaten by fish. The drainage basin of the Amazon is so vast that it has its own rainforest water cycle: so much water evaporates that it forms its own clouds and 'aerial rivers' which transport

Flood in the USA

water vapour to south-eastern Brazil and the country's two largest cities, São Paolo and Rio de Janeiro. After failed rainy seasons in 2013–14 and 2014–15, the water levels in main reservoirs fell to 5 per cent and less. At one point, the 22 million residents of São Paolo were down to less than twenty days' worth of water. Removing the trees reduces transpiration and cuts off the rain supply. For Dr Antonio Nobre, senior researcher at the National Institute of Amazonian Research, the connection is clear: 'In normal years, most of the rainfall feeding the south-east is carried from the Amazon through aerial rivers. The link of deforestation with reduced rainfall within the Amazon is well established. One only needs to connect the dots.' In the last fifty years, around 17 per cent of the Amazon's rainforest has been removed. Nobre sees 20–25 per cent as the tipping point, after which the Amazon system will 'flip to non-forest ecosystems in eastern, southern and central Amazonia'.

Pollution, whether from plastics, chemicals or general refuse, has become a universal problem for the planet's surface waters. One thousand years after Gerald of Wales – the father of Welsh geography – celebrated the River Teifi's salmon-dancing waters and beaver lodges and pools and falls, a slurry spill killed at least 1,000 fish in a 2-mile stretch of the Teifi. And in the last three years there have been almost 3,000 pollution incidents in Welsh rivers. Fish stocks are down, ecosystems are suffering. One of the major sources of pollution in the world's rivers is nitrogen from synthetic fertiliser, livestock waste, sewage and the burning of fossil fuels. Nitrogen flows have doubled in the last half-century. One of those working on this crisis is Xin Zhang, an environmental scientist at the University of Maryland. Her research found that the efficiency of nitrogen use has dropped since 1961 from over 50 per cent to around 42 per cent. More than half the nitrogen being applied to fields is washing into rivers. Already there are 400 'dead zones' in the oceans.

It is not in our interest to pollute the planet's natural waters. The 71 per cent of the planet's surface occupied by ocean is a vast reserve of biodiversity and a significant contributor to human food. The total annual marine catch currently fluctuates at around 80 million tons. And it is getting more difficult to supply markets. As coastal marine areas have been fished out, fisheries have expanded geographically, moving southward from the North Atlantic and North Pacific, and dropping their nets at 2,000 metres (from previous maximum depths of 500 metres). As traditional fish stocks disappear, unfamiliar species have been brought to the table, frequently rebranded to make them more appetising for consumers. The sluggish, deep-sea 'slimehead' became the 'orange roughy'. Worldwide catches of orange roughy peaked in 1990 and then plummeted. The true scale of the crisis in the oceans is not known. Around 1,500 fish stocks are commercially exploited, but comprehensive estimates of their health exist for only 500 of them. The knowledge gap gets worse when estimates are sought for over-exploitation: the Food and Agriculture Organisation of the UN puts the figure at 29.9 per cent, while the Hamburg-based World Ocean Review produced their own model and came up with a figure of 56.4 per cent.

Seawater is not what it was. Among the pollutants swilling in the oceans are oil, fertilisers, plastics, sewage and toxic chemicals. Oceans are also over-dosing on carbon dioxide. Around one-third of the carbon dioxide released through human activities has been absorbed in the oceans, increasing the acidity of the water to the detriment of marine organisms such as corals and shellfish. Scientists are concerned that acidification of the oceans will alter the competitive advantage between species and affect marine eco-systems as a whole. And this is a concern that again strikes global food supply.

Fixing the water cycle will not be easy. Resolving problems such as deforestation, pollution, plastics, nitrates and over-fishing will require radically new models of politics and business. Rising sea levels and the increased frequency and intensity of storms, heatwaves, droughts and floods will mean that we have to adapt as our habitats change. On the whole, the water cycle's components, processes and malfunctions are well known to science. But they are not well known in the wider world. The first step towards solving those problems is to deluge the world with droplets of knowledge.

THREE

Neuropolis

Like most of the planet's residents, I'm a city-dweller, a townie. And no matter how often I dream of green spaces, mountain peaks and crystal cataracts, I'm living in a London terrace with a concrete backyard. I'm very fortunate. In this chapter, I'm going to take a wander through the imperfect world of urban geography.

Babbanji ran away from home in rural Bihar and made his way to Mumbai, where he sleeps on the pavement and queues for a turn at the public toilet with his tote bag of poems. He appears on page 510 of *Maximum City*, Suketu Mehta's sweltering book about 'the biggest, fastest, richest city in India'. Mumbai is 'an island state of hope . . . a mass dream'. But it is also one of the most congested cities in Asia. Pavements are obstructed by sleepers, hawkers and vehicles. There is a neighbourhood that measures not much more than one square mile, packed with factories, tanneries, bakeries and sweatshops. And with around a million people, Dharavi is the most densely populated urban area on the planet. Why, asks Mehta, 'would you want to leave your brick house in the village with its two mango trees and its view of small hills in the east to come here?'.

For millions, life outside the city has become untenable. Rural poverty, lack of education, health care and entertainment, lack of opportunity and environmental crises like drought and flooding push people from field and home, while the city itself exerts a magnetic pull, promising better housing, jobs, opportunities and services; a better quality of life. Behind these pushes and pulls are a number of background factors: improved communication is increasing access to urban opportunities; the internet, radio and TV are spreading knowledge about urban employment. Changes to cities themselves are strengthening their attraction to rural outsiders. A potent dream lengthens the stride.

Babbanji is one of the majority on this planet who find themselves in a human maze. Cities were always dense, fluid and connected, but the ramping of numbers, flows and connectivity has forced them from their geographical niche. They have become the defining components of a world system.

The numbers tell an incredible tale. The 2 million who were striding the planet in the early centuries of this balmy interglacial swelled through millennia of foraging and hunting to around 18 million by the time agriculture was creeping across Europe in 5000 BCE. Already, there were cities in the Fertile Crescent. By 1000 CE, the world's population had increased to 295 million. Cities such as Kaifeng on the Yellow River and Baghdad on the Euphrates thronged with a million or so inhabitants, dwarfing by fifty times the European cities of Paris and London. China is thought to have had no less than five cities of a million people by 1100. By 1800, global population had crept up to 890 million, and by 1900 it was 1.6 billion. The twentieth century changed everything: by 1950 the figure stood at 2.5 billion, and by 2000 6.1 billion. As I write this, we're at 7.6 billion. Humanity's journey from hunter-gatherer to urban commuter has been accompanied by a 4,000-fold increase in numbers.

As populations grow, cities swell. As recently as 1950, two-thirds of the world's people lived in rural settlements. Today, 54 per cent of the human population live in urban areas, and by 2050 that figure will have risen to 70 per cent. These figures are accompanied by the spectacular emergence of the 'megacity', a conurbation that has grown beyond 10 million inhabitants. Back in 1950, the world had only two megacities: New

York and Tokyo. The UN predicts that by 2030 there will be forty-one megacities and another 662 cities with more than a million residents. China's cities have experienced such growth that many of them have congregated into city clusters, or urban mega-regions. On the Yangtze delta, Shanghai, Suzhou, Hangzhou, Wuxi, Ningbo and Changzhou have a combined GDP comparable to that of Italy. Two further mega-regions have sprawled across the delta of the Pearl River and around Beijing and Tianjin.

Seen from space at night, cities are the most visible geographical features on Earth. The Atlantic Seaboard Conurbation from Boston and New York to Philadelphia and Baltimore looks like a single lava flow. Chicago and its urban neighbours outshine the distant green aurora borealis. Magnified through a camera lens, Denver becomes a white-hot barbecue grid and Tokyo a blue-green amoeba (its colour is caused by mercury-vapour lighting). Europe is a garden of incendiary blooms.

Flowing to and from these hot nodes are hundreds of millions of people. Cities are more human than physical. The context of their existence is the long- and short-haul shuffling of *Homo sapiens*. Most of us are migrants; participants in the milling urban diaspora. Energy, the human geographer Danny Dorling has written, 'is the dynamic unifying everything, from plate tectonics and climate systems to the global economy and the culture of any place'.

Among the long-haul energy flows are the 258 million people (3.4 per cent of the world's population) who live beyond the borders of their own country, an incredible increase of 49 per cent since 2000. In numerical order, India, Mexico, Russia, China and Bangladesh are the greatest suppliers of expats, while the most popular destination for international migrants is the USA, which has become home to 19 per cent of the global total. Most choose to leave home because they are seeking work, or because they have been driven from their countries by persecution, natural disaster or war. A third category – asylum seekers – take to the road in order to apply for international protection. In the dislocated world of international migration, cities are primary magnets; they are the 'unaccustomed Earth' that Jhumpa Lahiri chose for the title of her immigrant stories. Pranab Kaku, who walks into a Lahiri story called 'Hell-Heaven', 'was so new to America that he took nothing for granted and doubted even the obvious'. Such questioning keeps cities – and civilisations – on the move.

Enormous numbers flow within countries, too. In India, the droves of people on the move have overtaken the country's capacity to provide reliable statistics. Estimates of the numbers of short-term 'seasonal' migrants (those who stay away from their usual place of residence for between one and six months) range from 15 million to 100 million. A vague estimate of 400 million has been placed on the number of internal migrants who have relocated permanently within India. Imagine the combined populations of the USA and UK, moving for work. Babbanji's story is played out on every city pavement; mass migration fed by a demographic surge and waves of poverty.

In terms of absolute numbers, no country matches the internal human flows of China. Following the establishment of the People's Republic in 1949 and the adoption of a Soviet-style growth strategy, rapid industrialisation was driven by increased agricultural output. The rural labour force was locked to the land through a *hukou* registration scheme that classified each person as 'rural' or 'urban' within a named administration unit. 'This is a war', Chairman Mao told his Politburo, 'on food producers – as well as on food consumers.' At breakneck speed, Mao aimed to convert China from a peasant economy into a military superpower. To pay for the 'Superpower Programme', the rural population would be educated to eat less, while the residents of towns and cities would be subject to rationing. The ligature was tightened in 1955 with the imposition of collective farms that prevented peasants siphoning food away from their own land. Much later, in the mid-1980s, economic reforms smashed apart the rural dams and released tidal waves of agricultural workers to pour into towns and cities, where they were needed by industry. 'This Great Migration,' wrote Kam Wing Chan in 2012, 'has supplied China with a mammoth army of low-cost human labor to power its economic machine.' Over three decades, between 200 and 250 million residents left their homes in the countryside and trekked to towns and cities, mainly on China's east coast. In terms of population shift, this far exceeded the 50 million or so Europeans who migrated to North America between 1800 and the First World War. China's fast-track industrialisation abruptly tipped the rural–urban balance. In 1980, around 80 per cent of China's population could be classified as rural, but by 2012 that figure had dropped to 50 per cent. By 2030, 70 per cent of China's population will live in urban areas.

Collectively, urban settlements have spread to girdle the Earth with connected streams of energy; they are hubs in a global network. Some have become 'world cities'. The prestige of a world city is built on economic power, proximity to regions of growth, inflowing foreign capital and political stability. And the emerging leader in this game is not Beijing or New York or London but Dubai, the new pole of accessibility, a place that has been described as the 'world's most future-forward city' and the 'centre of the world'. In

*Nearly three-quarters of China's population
will live in towns and cities by 2030*

recent years, the mall at the foot of the Burj Khalifa – the tallest building on Earth – has been the most visited place in the world. Dubai is a spinning hub, with a population that is over 90 per cent foreign-born and a fleet of Airbus 380s flying non-stop to every major city on the planet. In his book *Connectography: Mapping the Global Network Revolution*, Parag Khanna labels Dubai 'an experiment in catapulting from feudalism into postmodernity'. From a creek-side population of 20,000 in the 1930s, Dubai has grown to 3 million and is anticipated to reach 5 million by 2027. This is, writes Khanna, 'a new type of global city . . . with a new kind of identity, a truly global node whose virtue is not its rich cultural heritage but its stateless cosmopolitanism and seamless global connectivity'.

That connectivity extends to the gaps between cities. Nowhere habitable on the planet is beyond the reach of modern *Homo sapiens*, if not by 4x4s, boots or skis, then by drones and satellites. Even if we do not live on a city street, virtually all of us are dependent upon cities to provide the services, social systems, economic stability, security and governance upon which societies depend. This is a world away from the original, minority role of the cities that rose from the rich silts of Iraq, with their emboldening walls to mark the urban–rural boundary. As long ago as 1987, the chronicler of civilisation Fernand Braudel wrote that the 'West's first success was certainly the conquest of its countryside – its peasant "cultures" – by the towns'. Can it really be said that there remains a space known as 'countryside', in its authentic sense?

The clear human boundaries between Braudel's town and country have been erased by world-scale mutual dependencies. The blurring of urban boundaries is accentuated by new attitudes to green space. While the 'countryside' becomes city-dependent, cities are rediscovering their internal countrysides. Havana led the way with its micro-gardens or *organopónicos*, boxes and bins and halved oil drums wedged onto rooftops and into backyards and filled with organically grown vegetables. Under Mayor Sadiq Khan, London became the world's first 'national park city', a 1,572-square-kilometre tract of land that is 47 per cent 'physically green' and home to 14,000 species of wildlife. A Starbucks opens in Yosemite National Park, and a beaver brings down a tree in the city of Stockholm. In many parts of the world, the division between town and country is becoming more difficult to distinguish.

Such is the scale, dynamism and complexity of modern cities that a new unknown has been created. It used to be thought that to lose yourself you would have to take to the wilderness: plunge into jungle, desert or mountain range. Or take to the ocean in a small boat. But cities now offer a new kind of wilderness, unknowable, fascinating, ethnically startling and in many cases scarcely mapped. If you want to get lost, get on an urban bus or pull on a pair of trainers. Orientation can take time. In large cities short on historic or topographic landmarks, mental maps are slower to build. A new generation of explorers, cyclists and psychogeographers are remapping urban landscapes through non-places.

In this expanding urban maze, waymarks – as ever – are key to community cohesion. The swirling currents of humanity within the city move along their own social riverbeds and pools. Babbanji moved from his patch of pavement to the Sulabh Sauchalaya public toilet, to the Punjabi *dhabas* where he ate, to the bookstore where he worked, to the special places that inspired his poems: the site of a recently collapsed building; the lanes behind Flora Fountain where the drug pedlars sleep and trade; the Santa Cruz slum on its open sewer. Physical landmarks, both pristine and putrid, are the necessary waymarks on mental maps updated in 3D – with audio and odour – within our immensely versatile brains. Newcomers to cities have to build these maps quickly; they are the webs of subsistence. Far too often, city authorities demolish or obscure historic landmarks, often realising too late that they have destroyed the identity of a district and undermined the cohesion of its community. Hyderabad has sprawled across 650 square kilometres, but the minarets of Charminar remain the soul of the city. MANHATTAN. Even its capital letters look like skyscrapers. But lingering at the roots of New York's serene stalks is Central Park, scattered with glacial erratics and ice-polished schist that would have been familiar to the Lenni Lenape, the native Americans who named this island for its prized copse of hickory trees. In the Munsee language, 'the place where we get bows' is *manaháhtaan*. In other cities that have expanded too quickly for distinctive waymarks to have achieved the reverence of historicity, there can be strange stopgaps: the Eiffel Tower, Haussmann boulevards, half-timbered town houses and 'Thames Town' inserted into Chinese cities are placeholders until vernacular architecture recovers its confidence.

The rush towards urbanisation is on such a scale that in many instances the historic form and function of cities have been overrun. There was a time when the city was introduced to students as an apparently isolated, relatively static settlement that could be neatly subdivided into multiple nuclei, zones, sectors or districts. The Central Business District, the Medium Class Residential District and the Industrial Suburb were projected in lecture halls as if the internal borders of cities were demarcated with red lines on pavements. Cities were never this simple, of course, but the scale and pace of modern urbanisation and the diffusion of distinctive 'zones' have muddied and multiplied polarities and created an entirely new human landform.

Let me dwell for a moment on one example: a type of urban district that is fluid, often unmapped, numerically vast and known to its inhabitants by various labels, from *zopadpatti* to *favela, rookery* to *barrio*. In Dhaka it is a *boosti*, literally 'made of tin', and in Istanbul a *gecekondu*, from *gece*, meaning 'placed', and *kondu*, 'overnight'. The 'slum' is the world's fastest-growing human habitat and already home to one-quarter of the planet's urban population.

The term 'slum' is an unhelpful label. It has come to connote a 'lost place' rather than an integral part of the city. The word is also open to confusion: it has been applied through time to inner-city neighbourhoods in MEDCs (More Economically Developed Countries) where a district of formal housing has become informal, or run down, yet is still operated – more or less – by top–down government. It's not unusual for these kinds of districts to be gentrified by urban colonists, who turn them into sought-after 'villages'.

The kind of 'slum' found more often in LEDCs (Less Economically Developed Countries) usually develops as informal, makeshift dwellings, with aspirations to formalise and with some kind of bottom–up governance. The two types of 'slum' have entirely different roots and structures. Mike Collyer, whose 'Migrants on the Margins' research project is currently under way in cities in Sri Lanka, Bangladesh, Zimbabwe and Somaliland, prefers the term 'under-served settlement'.

The settlements Collyer and his colleagues are investigating vastly dwarf their MEDC cousins, in both scale and significance. There are many Dharavis, and they are the most fluid, densely occupied urban districts in the modern world. Constructed of low-cost materials, necessarily low-rise, with inadequate services and high rates of crime and disease, they are highly adaptive, self-repairing urban ecosystems which expand or contract according to demand. In the sense that they are a manifestation of market forces, they are typical of rapidly urbanising cities where civic governance and infrastructure simply cannot keep pace with demand from new arrivals. Under-served settlements may appear from the outside to be squalid warrens, but they are also hives of opportunists: recyclers of material, furniture-makers, bakers, hair-cutters, cobblers and seamstresses, food-sellers and phone-menders. They can be incubators of innovation.

This kind of creative friction was memorably described over fifty years ago by the great Jane Jacobs, who made her home in Manhattan, that circuit board of ambition wrapped by the shining solder of the Hudson. Jacobs was big on sidewalks and streets. She lived at 555 Hudson Street, and in her book *The Death and Life of Great American Cities* sidewalks and street corners are places where people meet and interact. Slums (she wrote of the 'formal' type of slum) were hotspots of diversity. Better than any writer of her age, she recognised that the success of cities was founded on density and diversity. 'Cities', she wrote, 'are an immense laboratory of trial and error, failure and success.'

Innovation is the urban USP. Cities are dense masses of hypermobile molecules generating round-the-clock intellectual friction. Innovative heat creates economies. And it can solve challenging problems. Measured by patent applications, as much as 90 per cent of innovation worldwide is generated in cities. Cities are power stations of the mind. They are the future of humanity.

Why are cities so effective? If we drop back in time and place to circa 2001 and the Santa Fe Institute in New Mexico, we meet the theoretical physicist Geoffrey West, who was beginning to explore with colleagues some ideas that would eventually turn into a 'science of cities'. West found that 'socioeconomic quantities' such as average wages, the number of professional people, the number of restaurants and GDP all scaled in a 'superlinear' fashion; for every doubling of population, there was an increase of 15 per cent or so per capita in these socioeconomic quantities. And this

Diversity in the Anthropocene

was true for innovation, too: the larger the city, 'the more innovative "social capital" is created'.

West's 'science of cities' also revealed that bigger can be more sustainable. Looking at European cities, his team discovered that the number of fuel stations increased with population size at a 'sublinear' scale of around 0.85. As West put it, 'with each doubling of population size, a city needs only 85 per cent more gas stations – and not twice as many as might naively be expected – so there is a systematic saving of about 15 per cent with each doubling. So, on a per capita basis, the large city needs only about half as many gas stations as the small one.' Other elements of urban infrastructure – such as the total length of roads, water pipes, electrical cables and so on – scale in the same way. Cheekily, West claimed that sublinear scaling made New York the 'greenest' city in the USA. But the model is not as clean as it might appear at first sight. West also uncovered a downside: that with every doubling of population, exposure to crime, pollution and disease also increases by 15 per cent per capita. And while superlinear scaling increases the amount the average citizen 'owns, produces, and consumes, whether it's goods, resources, or ideas' (thus explaining the attraction of cities to rural-urban migrants), a model that generates superlinear production and consumption is very bad news for young people.

Looking towards a decarbonised epoch, these glistening stalks within the green belt summarise our hopes. Cities have the capacity to innovate and become models of sustainability. In Copenhagen, an astonishing 62 per cent of the city's population cycle to work every day. In Stockholm, a project to use excess heat from a data centre has the potential to heat up to 10,000 households. In Australia, Melbourne's Council House 2 uses 'biomimicry' technologies to save water and energy. Colombia's second-largest city, Medellín, has led the way with 'library parks', ten cultural hubs surrounded by green space that have provided new social spaces for marginalised communities. In Brazil, the city of Curitiba has managed to decongest streets by providing a Bus Rapid Transport system that is used by 80 per cent of the population. Nobody in this city of nearly 2 million lives further than 400 metres from a bus stop.

Let's embrace these bulging urban brains, these massed flowerings of opportunity and potential. The frontiers of exploration have spread from deserts, forests and mountain ranges to include centres of human population whose increasing scale demands new research, new ideas, new policies; fresh symbolism. Urban jungles are greener than we think. The time has come to explore the neuropolis.

FOUR

Yü, Or How I Found My Inner Geographer

We all have a world view. It is one of the things that make us human. Wherever we come from, whoever we are, we all have to engage with people, places and our environment. Geography is a way of thinking. It has been with us since we broke from our drying forests and took to the savannah, walking and running on two legs, using sticks and stones as tools, working together in social groups, domesticating fire to manage landscapes, labelling places and compiling mental maps.

Back in the spacious dawn of hunting and gathering, our aggregated behaviour was insufficient to interfere on a global scale with Earth and its systems. But that is no longer the case. Our appetite for the planet is so all-consuming that we are eating it to death. We are geographical gluttons. The aptitudes that have guided us along our unique evolutionary path – allowing a tailless ape to build cities – have also provided us with the wherewithal to ransack our own life systems. Yet the geographical circuitry that trashed the planet also informs the algorithms of our future. We need to reorientate our inner geographer.

None of us are far from our inner geographer. We're all hard-wired with an extraordinary menu of spatial aptitudes. They can be detected during a location-changing lunch break or holiday. They can be refined through life. My closest encounters with them came during a long, solitary journey at the end of the last millennium. For one and a half years, I walked alone along the mountain watershed of Europe, from one side of the continent to the other; from the Atlantic Ocean to the Black Sea. I had no cellphone or GPS, and for most of the 10,000 kilometres I slept beneath the stars. As the ranges unrolled and seasons turned, I grew less reliant on printed maps and more adept at my own, neural, surveying. And I was compiling cognitive maps of the most extraordinary kind. They were 3D and at any scale I chose, even 1:1. They logged sound (the rumble of cataracts, the crack of lightning, the stirring of wind in pines) and touch (the feel of angular limestone under foot, the caress of grass) and smell (the chill tang of rain or whiff of sheep, and implied presence of ferocious dogs). They recorded relationships: the way rivers and spurs interacted with gradients, the action of weathering in excavating cols, the give-and-take between geology and paths, the seasonal rhythms of shepherds and flocks between valley and upland. Without the ability to compile my own models and to compute interactions between topography and weather, between gradient and traction, between river width and depth and so on, I would have been dead in a week. My imagination was a multisensory fieldworker, absorbing and encoding data that could be retrieved at a moment's notice. My well-being depended upon the intuitive understanding of scrolling geographical space. This auto-updating came naturally, as if my senses and memory were hardwired to analyse and record the passing topography. In the words of the geographer Kent Mathewson, I was exhibiting 'universal traits expressed as geographical competencies'. We can all do it. And we start young.

An insight into universal geographical competencies was provided in 2003 by a quartet of geographers and psychologists in the USA and UK. James Blaut, David Stea, Christopher Spencer and Mark Blades hypothesised that 'nearly all humans, in all cultures, acquire the ability to read and use "map-like models" in very early childhood'. We should pause a moment to define a map-like model. In this context, it is not the same as a conventional 'map', which demands that its user be able to read words and understand cartographic conventions that must be learned. James Blaut in Chicago and his psycho-geographical colleagues on Chautauqua Hill and the River Don defined a map-like model as the kind of descriptive device that might be created and used by non-literate cultures or young children. It is abstract, in that the information on it is a much reduced – and perhaps distorted – version of the real landscape. Their studies showed that 'children everywhere, perhaps by their fourth birthday, can deal with map-like models', and that this is preceded by a period of 'pre-mapping' abilities that begins at birth.

Not only are map-like models the spatial tools of our formative years, but they are also cultural universals, methods of thinking and acting that are common to all cultures. Just as the use of language, tools, shelter and food can be identified as universals, the use of map-like models by young children can be tracked from continent to continent. The ability to use them is the beginning of the geographer in all of us.

My ability to create and process my own cognitive geographies while threading the passes and crests of Europe's watershed was a response to spatial need: I was lost without them. The landscapes I was navigating were 'macro-environments' or 'geographical spaces'. They were too huge for me to grasp at a glance from any one viewpoint and were quite different from the object-scale 'micro-environments' I'd meet during a comfort break or bivouac, when I'd quickly familiarise myself with – and mentally map – prominent features like rocks or trees or a stream. But a macro-environment is so large that the only means we have of viewing it holistically is by imagining it as a model. And we've been doing this since *Homo habilis* took off across the drying savannah in search of a tasty carcass. As Blaut et al. suggest, 'mapping, or map-like modeling, is a basic and necessary part of the strategy used by nearly all people, in almost all cultures, to cope successfully with the macro-environment'. From the youngest age and from the earliest times, humans have been geographical beings. 'Being geographical is inescapable,' wrote Robert Sack in *Homo Geographicus*, 'we do not have to be conscious of it.'

Our inner geographer has been expressing itself since the Palaeolithic. Although we have no way of knowing exactly what was meant by many of the lines and shapes in prehistoric 'rock art', every one of them is geographical in the sense that its location – fixed to rock or bone – is a commentary on 'place'. The cave bears, mammoths and rhinos lurking in the shadowed recesses of the Chauvet cave in France are the ghosts of lost ecosystems, the wraiths of landscapes gone. They hint at the biogeography – the life-form distribution – of 30,000 years ago. Similarly, the depiction of human figures or anthropomorphs on a rock wall illustrates a relationship that once existed between the people of that time and that location. Rock art can be seen as an expression of 'topophilia', the term provided in 1974 by the geographer Yi-Fu Tuan as he constructed his framework for discussing the various ways in which we develop 'a love of place'. For Tuan, a place could be personally scaled, from a house to planet Earth. It was a 'field of care', a notion wrapped in belonging and value. 'Geography', he concluded, 'provides the content of topophilic sentiment.' And there are practical reasons for nurturing these spatial attachments.

Before the development of writing, landscape was memory. For millennia it has been understood by spatially sensitive humans that information learned in a given environment is best recalled in that same environment. Despite the efforts of psychologists from the 1930s onward, empirical evidence for 'context-dependent memory' was lacking. But in 1975, Duncan Godden and Alan Baddeley at the University of Stirling revealed in the *British Journal of Psychology* the findings of experiments they'd undertaken in two differing natural environments: under water and on land. Results showed that the recall of their subjects was better when the environment of original learning was reinstated. In a nutshell, human memory increased in utility when attached to a familiar place.

The many applications of context-dependent

memory go some way towards explaining why human beings have developed deep attachments to place. And these attachments would have existed long before we lived in permanent, year-round settlements. Groups of hunter-gatherers used landmarks such as rocks, trees and rivers to upload information for later retrieval. Records of births and deaths, the whereabouts of food and water sources and the boundaries of tribal or familial areas were filed in the mind, but retrieved through connections with the landscape. These memory banks are also likely to have been the repositories of human narratives, a subject explored in 1994 by the archaeologist/anthropologist Christopher Tilley:

> If stories are linked with regularly repeated spatial practices, they become mutually supportive, and when a story becomes sedimented into the landscape, the story and the place dialectically help to construct and reproduce each other. Places help to recall stories that are associated with them, and places only exist (as named locales) by virtue of their emplotment in a narrative.

In a paper published a couple of years later, Tilley went on to suggest that places such as the eye-catching granite tors on Bodmin Moor in south-west Britain would have been suitable locations for retrieving information:

'non-domesticated "megaliths" or stone monuments, sculptured by the elements and imbued with cultural significance in the Mesolithic imagination in the forms of stories, myths and events of cosmological import'. The massive Okotoks erratic that I described earlier was embedded in the legends of the Blackfoot tribe. More recently, it gave its anglicised name – Big Rock – to an Albertan brewery.

Maybe we should see our hunter-gatherer ancestors as being better connected than urbanised *Homo sapiens*. Rock art was an integrated experience in which the viewer became a participant, linked to the natural world. The surfaces of the bedrock, its clefts and bulges, were frequently incorporated within the composition. Touching them brought the observer into the physical realm of the spirit world. Landscape, spirit beings and artist were as one. Researchers have suggested that the 'consumption' of rock images required altered states of consciousness, synaesthesia perhaps, accompanied by a transference in which the viewer became the image. A modern, mechanistic equivalent might be the use of avatars in virtual-reality geographies.

The closest we can come to the geographies of our prehistoric ancestors is by looking at the worlds of hunter-gatherer groups who survived into the modern age. For the San of Nomansland below the Drakensberg Escarpment in South Africa, and for other groups too, there was more than one world. Connected to the world of humans was a supernatural one populated by spirit people and gods. Two-way traffic was possible, frequently via waterholes, which were often the focus for concepts of territory, areas of natural resources bounded by major landmarks. Not surprisingly, the San were preoccupied with rain; their rock art frequently included 'rain animals' which had to be lured to waterholes by shamans. An element of social geography was portrayed on some of the Nomansland sites, where portraits of powerful 'potency-owners' who may have 'owned' the site also controlled the creation and consumption of images.

We see the same kind of interconnected worlds on the northern fringes of Norway, Sweden, Finland and the Kola Peninsula of Russia, where Sámi rock-art sites are thought to date back over 4,000 years. The Sámi were hunters, fishers and gatherers who believed that everything – including the stones and trees – was animated and that the origin of all life was the Earth Mother, Máttaráhkká. Specific topographic features had meanings: mountains might be the homes of ancestors; rapids, caves, ravines and peaks could be gateways to other worlds. Small lakes with no outflows or inflows were believed to be underwater portals to the lower world. Máttaráhkká's abode was the holy mountain of Áhkká, a massif of a dozen peaks and

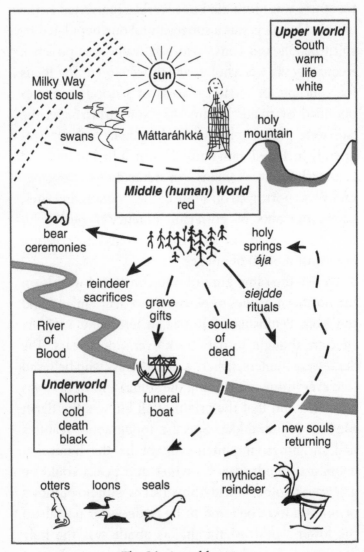

The Sámi world system

almost as many glaciers, on the edge of what is now the Sarek National Park. At Sámi art sites, the surface of the rock itself was sacred and a part of the image, as was its aspect and location.

The Sámi cosmos had three tiers, with humanity at the centre, located between the upper world and the underworld. A world-tree or pillar connected all three. When the Swedish archaeologist Inga-Maria Mulk and the Cambridge geographer Tim Bayliss-Smith represented this triptych graphically, the warm, white, southern upper world held the Sun, holy mountain and Máttaráhkká; the red, middle, human world had a congregation of stick figures, holy springs and bear ceremonies; and the cold, black, northern underworld was populated by loons, otters, seals and mythical reindeer. The imagined geographies of the San and Sámi were interconnected spiritual systems complete with feedback loops, and there were implications should connections be broken.

Geography also underpins the *Epic of Gilgamesh*, in which Uruk's hapless king journeys to the ends of the earth in search of the secret of immortality. The poem's thirteen 'mighty gale winds' summoned to bash the dreadful Humbaba read like a meteorological checklist: 'South Wind, North Wind, East Wind and West Wind, / Blast, Counterblast, Typhoon, Hurricane and Tempest, / Devil-Wind, Frost Wind, Gale and

Tornado'. These were weather conditions known to the Sumerian consumers of the epic. They were recognisable geographical forces. It's also a journey between geographical extremes: Humbaba's mountainous 'Forest of Cedar' is the distantly wild counterpoint to the civilised, secure city of Uruk, with its magnificent ramparts of brick, 'sevenfold gates' and Temple of Ishtar. Three thousand years ago, 'Uruk-the-Town-Square' was the place where a poet could review the past and weave its stories, whereas the space beyond was for journeys: north to the Cedar Forest and its psychopathic monster, Humbaba; west towards Death. Like the Sámi and the San, the Sumerians of the Euphrates lived in a connected, supernatural world. Uruk's presiding triad was Anu the Sky God, Ea of the Ocean Below and Enlil, who governed the affairs of men and gods from his temple on Earth.

The record of geographical understanding solidifies in 1900 BCE, where a shadowy sage-king called Yü harnessed – according to legend – the disruptive waters of the Chinese heartland and created an agricultural superpower. No contemporary accounts of Yü exist and he doesn't appear in inscriptions or on oracle bones. The main source for his heroic endeavours is the *Yü kung*, the *Tribute of Yü*, which can be found in China's oldest complete classical work, the *Shu Ching, Book of History*, of around the fifth century BCE.

Embedded in the *Tribute* are geographical descriptions. The habitable lands of the ancient Chinese were located 'within the four seas'. Their imperial capital occupied the centre of a system of concentric zones: a nest of human geography in which the metropolitan zone was surrounded by the Royal Domains and then the Domain of the Nobles, with its cities and lands of high ministers, officers, barons and princes. Beyond that stretched the pacification zone, a Marches borderland where Chinese civilisation was being adopted through 'lessons of learning and moral duties' and 'the energies of war and defence'. Further out lay the zone of allied barbarians and criminals 'undergoing lesser banishment'. At the outermost edge was the zone of 'cultureless savagery' where the tribes of Man and 'criminals undergoing greater banishment' could be found. (The *Tribute*'s enveloping zones bear a striking similarity to the concentric-zone model developed for American cities in the 1920s.)

Physical geography was recorded in the *Tribute* through the description of nine provinces bounded by topographic features such as mountain ranges and rivers. Also named were five political domains, more than thirty-five rivers and their courses, and forty-five mountains and hills. And the *Tribute* identified the geographical feature that formed the cradle of Chinese civilisation. For millennia, deposits had accumulated

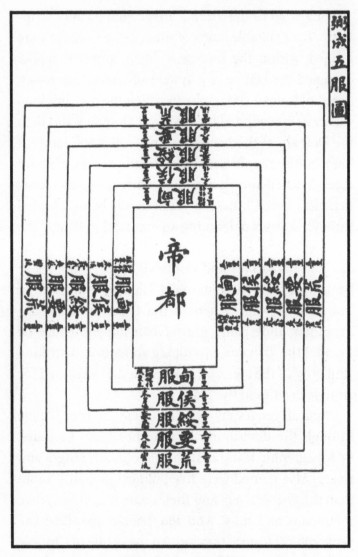

The geography of Yü

in the northern part of central China to form a vast plateau of some 620,000 square kilometres. Known today as loess, this fine-grained, loose-textured deposit is high in mineral nutrients and proved a productive bed for arable agriculture. In places, it lay over 150 metres deep. In the *Tribute* the shifting soils of loess were called the 'Moving Sands'. Across the loess flowed the mighty, flood-prone Ho, later known as the Yellow River, after the prodigious amounts of pale sediments borne by its currents.

It was Yü's efforts to geo-form the flood-prone lands of China that elevated him to legendary status. Aided by armies of imperial labourers, he felled access routes through the forests, channelled, ditched and dyked the great rivers and drained marshes and converted others into lakes:

The waters of the Heng and Wei were brought to their proper channels and the plain of Ta-lü was made capable of cultivation ... The nine branches of the Ho were made to keep their proper channels. Lei-hsia was formed into a marsh in which the waters of the Yung and the Tsü united. The mulberry grounds were made fit for silkworms; then the people came down from the heights, and occupied the land below ...

And so Yü labours on through the *Tribute*, draining, banking, diverting and planting. The Rivers Wei and Tsi were embanked within their original channels; the 'wild people' of Lai were taught tillage and pasturage; the hills of Meng and Yü were brought under cultivation; the lake of P'eng-li was confined to its shores so that wild geese could settle; and the region of San-wei was made habitable:

> Thus throughout nine provinces order was effected: lands along the waters were everywhere made habitable; hills were cleared of their superfluous wood and sacrificed to; sources of rivers were cleared; marshes were well banked; access to the capital was secured for all within the four seas.

Yü became known as the 'Tamer of the Flood', founder of China's first dynasty, the Xia. But is there a verifiable link between the 'Tamer of the Flood' and a known flood event? In 2016, a team of Chinese scientists sought to connect the hydraulic endeavours of Yü to catastrophic flooding that followed the collapse of an earthquake-triggered landslide dam. The geologist Qinglong Wu of the China Earthquake Administration in Beijing had been conducting research on the Yellow River when his team came across sediments from an ancient lake in the Jishi Gorge. Wu and his team concluded that a landslide had dammed the gorge, causing a huge lake to rise behind a wall of rubble, which had eventually given way, releasing between 11 and 16 cubic kilometres of water, so much, they claimed in their paper, that the flood 'could easily have travelled more than 2,000 kilometres downstream'. They dated their Jishi Gorge catastrophe to around 1900 BCE, when Chinese society was passing from the Neolithic into the Bronze Age and when the Erlitou culture was transforming the landscape with palace buildings and smelters 2,500 kilometres downstream. To Wu and his colleagues, this was 'an illustration of a profound and complicated cultural response to an extreme natural disaster that connected many groups living along the Yellow River'.

Wu's paper unleashed a digital flood of controversy, and for the time being Yü's association with recorded floods and the founding of the Xia dynasty have to be left open. Meanwhile, the *Tribute* itself has attained a life of its own. Revered in China for over 2,000 years, it was translated into English in 1865 by the missionary-sinologist James Legge, a Scottish Marco Polo who devoted half a century to bringing Chinese classics to the Western world. In 1959 the *Tribute* was lauded by the great sinologist Joseph Needham in his multi-volume *Science and Civilization in China* as 'the first naturalistic geographical survey in Chinese history'.

By 1984, the Geographical Society of China in Beijing was promoting the *Tribute of Yü* as China's 'first geographical document'.

Regardless of Yü's engineering prowess, what we have in the *Tribute* – written over a millennium after Yü's hydraulic excursions – is the emergence from imagination, myth and legend of a document recording geography in its many facets, from physical to human, together with refracted connections from social and economic to political. Behind every civilisation is an understanding of people, places and the environment. Reading between the lines of the *Tribute*, it's difficult not to see Yü as the world's first recorded geographer. In China, he is Da Yü, Great Yü. Perhaps he should be the Great Geographer.

FIVE

One to One

Geography has many means of expression, from written descriptions to graphics, speech and gigantic digital memories stuffed with geospatial data. But the means most associated with geography is the map. Ubiquitous, versatile, timeless, all-encompassing, the map *is* geography. In this chapter, I'm going to look at the innate, universal nature of mapping and suggest that it is key to resolving our current challenges.

'If a man intends to visit a country little known to him,' wrote Dr Franz Boas in *The Central Eskimo*, 'he has a map drawn in the snow by someone well acquainted there.' In his book of 1888 the pioneering anthropologist describes how the mapmaker would begin by marking key points that were well known and then inserting the details. These snow-maps could be incredibly accurate.

Wherever they are on the globe, people can share

spatial information through the medium of simple maps. My own notebooks are sprinkled with sketch maps describing routes up mountains, through forests and across deserts and canyons, reminders of where I've been and, in some instances, fallback plans should I have to reverse my route. Some were drawn in order to ask directions. I remember my cousin Richard making a map in the grit of the Tibetan Plateau. We were crossing Central Asia on bicycles and had come to a parting of the ways, not far from a cluster of nomad tents. Using fingers to draw roads and stones to mark settlements, a cyclist and a yak herder found a common cartographic language.

At the other cartographic extreme, maps are complex systems of communication expressed through logograms, symbols representing geographical features. Augmented with schemes of colouring, shading, grids, scales and so on, complex, formal maps become miniaturised versions of reality. Differing traditions, resources and markets have created a wide diversity of cartographic styles. If you've travelled on foot through foreign lands using large-scale maps, you'll be familiar with the idiosyncrasies of national mapping agencies. Mount Iwatakeishiyama, Snowdon and Mount Whitney are all mountains, but the graphic tools used to describe them vary widely.

In short, maps come in many forms, from crude

sketches to fantastically intricate mathematical models. Maps are the endoskeleton of geographical knowledge. And cartography is careering through a revolution so dramatic that we have to rethink the word's original meaning. We have at our fingertips a new means of understanding our planet, and as we tackle an extraordinary range of global issues in the coming decades, it will be one of our most crucial tools. So, let's start with a bit of context.

We are each at the centre of our own map. You are at the centre of a mental map surrounding your home. Through its choice of culturally derived cartographic conventions, the Ordnance Survey is at the centre of its maps of Britain. In the Iraq of 4,000 years ago, Gudea, Prince of Lagash, was at the centre of his map, a plan he authorised – perhaps even devised himself – of a temple dedicated to the great god of Lagash, Ningursu. The plan was inscribed onto a tablet of polished igneous stone. Today, the tablet can be seen resting on Gudea's lap in the Richelieu wing of the Louvre in Paris. Gudea has lost his head, but his tautly erect torso perches on his diorite stool as if he's a passenger on a crowded Métro train, tablet on his thighs. He appears to be contemplating his latest architectural concept, his temple, his world. Poised on the tablet are a stylus and a graduated ruler, the tools of piety. The Prince of Lagash commissioned this stone selfie when southern

Mesopotamia was dotted with cities and independent dynasties, long before the people of America and Europe had begun to congregate in any place resembling a town. Iraq was the seedbed of civilisation, the Euphrates and Tigris channels of innovation.

From architectural plan to city plan. Six hundred years on, Lagash and the lower Euphrates had become part of the Babylonian Empire and, for reasons that are lost in the dust of crushed tablets, in around 1500 BCE somebody mapped the outline of a city then standing on the plain of southern Mesopotamia. The remains of Nippur exist to this day in the desert south of Baghdad, a desiccated mound of debris almost a mile across. First settled in about 5000 BCE, the site evolved into the most important religious centre in Mesopotamia, the great seat of the worship of Enlil, lord of the Earth. On a broken tablet now archived in Germany's Friedrich-Schiller-Universität Jena, the city of Nippur is defined by encircling walls that enclose the essentials of urban life: waterways, storehouses and temples. The Euphrates can be seen flowing beside the walls, which are pierced by named gates. It is a remarkable cartographic image, compiled perhaps to aid the reconstruction of defensive works.

If we linger on the Euphrates for another few hundred years, we see the mapmaker's field of view widen yet further, from city to planet. A tablet in Gallery 55 of

Nippur in Iraq

the British Museum shows a circular Babylonian world with the Euphrates flowing through its heart. The tablet, of unbaked clay, was probably made in Babylon in the seventh or sixth century BCE but may have been copied from a far older ninth-century version. Lines of text on both its sides describe places distant from Babylon: ruined cities, a desert or mountain range beyond the reach of winged birds, a region of trees, a part of the Far East where the Sun rises. There are animals of faraway ecosystems: a gazelle, lion, wolf, monkey, ostrich, chameleon and more. The map occupies about two-thirds of one side of the tablet and makes no attempt to represent true scale. The world is a disc, surrounded by a ring of water labelled 'Bitter River'. Topographic features include an 'outflow', a 'marsh', a 'mountain', kingdoms and tribal territories. A waterway (possibly a canal) may be an antecedent of the Shatt-al-Arab. In another echo of modern practice, the central city on the map, Babylon, is marked with a large label, while smaller, outlying centres are shown with a circle or a dot. The historical geographer Dr Catherine Delano Smith has pointed out that the cartographic conventions being deployed by the Babylonians point to 'established conceptual and mapping traditions'. They'd been making maps for some time.

In this trio of maps – a temple, a city and the world – we see three geographical models at varying scales.

That these models were crafted in the same part of the world over a period of 1,500 years or so is an extraordinary testament to Mesopotamian continuity; the Euphrates was a geographical hotspot. A prerequisite for complex geographical modelling – and indeed, all forms of advanced enquiry – was intellectual sanctuary. The great civilisations that had been nurtured by the waters of the Euphrates and Tigris produced a broad spectrum of spatial awareness. The place we know today as Iraq offered security, patronage and surpluses of time that could be devoted to the exploration of concepts of space.

Mesopotamia was not the only hub of geographical enquiry. On the Nile, the land of Egypt had been united under the authority of a single ruler since around 3100 BCE, and the civilisation that prospered upstream of the river's outstretched delta developed its own traditions of spatial expression. The plans and maps of ancient Egypt lacked the sophistication of those of Babylonia, but sufficient examples have survived to illustrate that modelling parts of the world in 2D was an effective method of communicating traditions and ideas. From at least 2000 BCE, coffins were painted with imaginary lands (with blue for water and black for an overland route) through which the deceased would travel. At Thebes, tombs were decorated with map-like diagrams of idealised gardens: date-lined paths, bodies of water, walled orchards and stands of sycamore.

The earliest geographical document to have survived from ancient Egypt was removed in around 1820 from a family tomb in Deir el-Medina on the west bank of the Nile by an Italian antique collector, Bernardino Drovetti. The 'Turin Map' shows a region of hills, wadis and roads, sketched onto a fragmented 2.82-metre papyrus scroll dating from the reign of Rameses IV (1151–1145 BCE) and annotated in the everyday cursive script of the time. Although the scroll had aroused the curiosity of scholars in the 1840s, it was not until 1914 that the Egyptologist Alan Gardiner revealed in the *Cairo Scientific Journal* that the black and pink hills on the map represented different types of rock. Finally, in 1992, after fieldwork on the Nile, two geologists from the University of Toledo in Ohio published their startling reinterpretation. The papyrus fragments became 'The World's Oldest Surviving Geological Map'. The brownish-pink on the scroll represented gold-bearing metamorphic and igneous rocks. Sedimentary rocks were coloured a dark brownish-grey. A pink hill with radiating brown stripes correlated with a granite hill that is still veined with iron-stained, gold-bearing quartz. A cistern marked on the map may have held water being used to separate gold from pulverised quartz. All this tied in with the text references to 'mountains of gold', 'the gold-working settlement' and 'the place in which they work in the great business of bekhen-stone which was established as a quarry'. Compiled when Rameses IV was conducting quarrying expeditions to Wadi Hammamat, the map recorded the locations of prized stone quarries, but it also described the general topography of the area, including communications.

Another cartographic sanctuary serves to illustrate the extraordinary power of maps to facilitate change, as well as the way that cartographic ideas are transmitted. At around the time that Babylonians were modelling their impression of the planet, spatially aware thinkers were also congregating on the coast of modern Turkey, at a place where the forces of physical geography had combined to produce a perfect haven: a warm gulf fed by the 'flood deep rolling' river we met in an earlier chapter. In 620 BCE, the Maeander flowed into the clear-watered Gulf of Latmos on the fret-worked eastern fringe of the Aegean Sea, convenient for coastal traffic plying the trade routes between the Mediterranean and the Black Sea. The gulf was protected from sea winds by a pair of peninsulas, an island and encircling mountains. Earlier settlers included Carians, who had probably migrated from islands in the Aegean and given their name to the region, Caria. A two-hour walk south of the gulf, the oracle of Apollo resided at the village of Didyma.

On the gulf's southern shore rose the city of Miletus. The blue lagoon of Latmos, with its sheltered anchorage, sea access, river-mouth , sheep pastures and oracle over the hill, was all that a civilised Greek desired. By the early part of the seventh century BCE, Miletus was the hub of a maritime state that had founded upward of forty-five – some said ninety – colonies along the shores of the Mediterranean and Black Sea. And it was home to the Milesian 'school', a troupe of brilliant minds devoted to investigating the universe.

According to Diogenes (and later, Aristotle), the founder of the school at Miletus was Thales. From Diogenes we also learn that Thales had a follower, Anaximander. Between them they investigated the nature of the cosmos, developing diverse and opposing theories. At this balmy campus in the eastern Mediterranean, the ideal of dialogue and open critique was pursued and science was born. Anaximander – who was critical of his master, Thales – came up with the astonishing theory that Earth was free-floating and held in place because it was equidistant from the edges of the universe (an equilibrium of forces not unlike those at the place known as L1, which opens this book). Modern thinkers have extracted Anaximander from classical entombment and wreathed him in laurel. His argument that Earth is fixed because it is equidistant from all other parts of the universe was lauded by Karl

Popper as 'one of the boldest, most revolutionary, and most portentous ideas in the whole history of human thought'. Patricia O'Grady, biographer of Thales, rated Anaximander's theory 'a brilliant hypothesis'. More recently, the theoretical physicist Carlo Rovelli described Anaximander's argument as 'extraordinary and perfectly correct'. Anaximander also produced a map.

'He was the first to draw a map of the earth', claimed Diogenes, 'and the sea and he also constructed a sphere.' The map itself has not survived, but several experts have attempted to recreate its form using fragments of description from those who saw it – or who heard about it. Anaximander thought that the world was shaped like one of the drums of a stone column, with the habitable part on the flat upper surface. His world was therefore circular. But what place marked the centre? Some have argued for Egypt, others for Miletus and others for Delphi. More certain are the various parts of the map Anaximander displayed. It showed the continents of Europe, Asia and Africa, various lands and seas, and the surrounding 'Ocean' that reached to the edge of the world. Compared to the Babylonian world map – created at about the same time – Anaximander's world was in a different league. It was a geographical map, and the first of its kind. He created – according to the Anaximander specialist Dirk Couprie – 'a new paradigm in mapmaking'.

The Milesian school flourished on connectivity. The city's own creation was a diaspora story. Thales was a spatial adventurer, sailing the eastern Mediterranean, dealing in commerce and breaking free from geographical explanations founded in myth, and turning instead to mathematics. He measured the location of places on the Earth's surface and used Egyptian geometry to solve geodetic questions. Anaximander brought the Babylonian gnomon to Greek science, using the sundial to calculate diurnal time and solstices. A third Milesian geographer, Hecataeus, used observations he'd harvested while travelling the lands of the Black Sea and Asia Minor, Egypt and Greece to compile his two-volume *Periodos Ges*, 'Circuit of the Earth', a pioneering work that described topography and ethnography alongside mythology and travel information. Sea-lanes and rivers were the fibre-optic cables of the ancient world.

Mapmaking reached its classical apogee with an Alexandrian librarian, Claudius Ptolemy, whose eight-volume *Geographia* included a huge gazetteer with locations tabled in longitude and latitude. The final volume was taken up with maps of parts of the world, and he also produced a world map. So exhaustive and accurate were Ptolemy's locational tables that mapmakers were using them nearly 1,500 years later as the European Renaissance gathered momentum and surged northward over the Alps towards Germany and the Low Countries, where mathematical mapmaking would be resurrected by humanists such as Gemma Frisius and Gerard Mercator.

The map story that meanders from the Euphrates to the Nile, the Arno and Po, the Rhine and the Schelde is the one that has directed Western minds. But there is a parallel narrative further east, in China.

In the story of China's dynastic birth pangs, the Xia and the Shang were followed by the Zhou, who managed, more or less, to hang on to power from 1100 until 221 BCE. Their denouement came with the Period of the Warring States, and it was in the midst of these troubles that a tiny state, Zhongshan – it means 'central mountains' – elevated itself to geographical immortality.

These days, Zhongshan is a playground for Beijing, offering peaks and gorges within reach of the capital and boasting on YouTube of its world-beating 488-metre glass-bottom bridge, a filament of wire and glazing slung between two cliffs. Zhongshan has a tradition of punching above its weight; one of its kings was once accorded the title of *wang*, a privilege usually reserved for the Zhou ruler. And in the realm of map history, Zhongshan is up with the best.

It was King Cuo of Zhongshan who was interred with a bronze plate bearing the plan of five sacrificial halls, four other buildings and two perimeter walls. The *zhaoyu tu* (mausoleum plan or map) is annotated with information that includes a royal decree, and dimensions of the buildings and intervening distances. It is the earliest known Chinese example of a bird's-eye topographic view, and the measurements (in *chi*, a foot, and *bu*, six paces) have led to the claim that this is the world's oldest map clearly marked with distances. The authority on material power in ancient China, Xiaolong Wu, has argued that bronze artefacts were used by their patrons 'in the assertion, negotiation, and communication of identity and power in their political and personal lives'. The map in Cuo's tomb was more than a knick-knack.

Besides their role as spatial tools, Chinese maps were compiled to demonstrate power, to educate and to please the eye. The four sodden wooden boards that were found in a tomb in the foothills of the Qin Ling Mountains have become a cornerstone of China's cartographic story. The Qin Ling are the strategically important range of uplands and gorges that separate northern from southern China; the tomb was one of a hundred or so and it contained the remains of a man named Dan. The boards that accompanied Dan to the afterlife were about one centimetre thick and covered in faint lines and annotations. After two years of controlled drying, a set of seven maps became clear enough to interpret. The first scholar to describe the maps, He Shuangquan, used records on eight excavated bamboo slips to deduce that Dan was a military officer who had committed suicide after injuring someone in the face; subsequent scholars such as Xiugui Zhang and Mei-Ling Hsu took the view that Dan was a scholarly civilian official. Collectively, the maps show an administrative district of particular importance to the Qin state of 300 BCE: the valley of the River Wei and its tributaries, together with a portion of the Qin Ling range and a defensive pass that connected the heartland of the state with the lands to the west. Black lines represent rivers and valleys, squares settlements. Labels identify transport checkpoints, passes and a variety of trees, from pine and fir to cedar and orange. The forest information lifts the map beyond mere cartography into the realms of economic or resource mapping. With its information about the location of timber suitable for felling, the Qin map has a comparable function to the 'Turin' geological map of Wadi Hammamat. Both inform the authorities of the time how to extract the best from their territory.

One final example illustrates the eminence of Chinese mapmaking. Until it was excavated, Tomb No. 3 was a bump on a small hill known as Ma-wang-Tui – 'mound of the Horse King' – on the outskirts of the city of Changsha in southern China. In 1973, engineering works for a new hospital brought archaeologists to the site and over 1,000 items were discovered, among them the world's oldest sex handbooks and three maps drawn on silk. The incumbent of the tomb was a man in his thirties who died in 168 BCE. Mei-Ling Hsu, whose research took her deep into the context of the maps (and the Qin maps, too), rated them for their 'extraordinary quality' and for their 'scale consistency, information content, and use of symbols'. She was restricted to examining the two maps that had been restored, one topographic and the other military. The square topographic map covered the south-central part of the modern-day Hunan Province and some adjacent areas. Streams and mountains are shown, and elements of human geography such as roads and settlements. The military map is a large-scale blow-up of part of the topographic map, with the addition of colours to emphasise significance. Military installations and headquarters are shown in bold colour. Intriguingly, Hsu was convinced that the symbol used for hills 'can be interpreted as expressing the primitive but basic notion of contouring'. If so, it would put the maps from the Mound of the Horse King nearly 2,000 years ahead of the conventional pioneer of contouring, the British mathematician Charles Hutton.

Three hundred years before Ptolemy produced his great works of cartography, it is clear that Chinese mapmakers were already proficient at many of the principles of modern mapmaking. In an essay he contributed to the monumental *History of Cartography*, Cordell Yee argued convincingly that 'It is not that Chinese mapping was non-mathematical, it was more than mathematical.'

The Ptolemy of ancient China lived between 223 and 271 CE. Pei Xiu broadened the scope of Chinese cartography from the descriptive to the analytical by declaring that six principles were to be followed: scale should be defined with graduated lines; a grid should be deployed for locational reference; and right-angled triangles used for calculating distance. The remaining three principles were concerned with transferring measurements made on the uneven surface of Earth to the flat surface of a map through conversion of elevation, direction and gradient.

It was Pei Xiu who used a map to resurrect the Great Yü, the founder of geographical description in China. One of two extraordinary maps Pei Xiu produced was known as the 'Yu gong diyu tu' ('Map of the Lands According to the "Tributes of Yü"'). Sadly, it has not survived. And neither has its partner, the 'Fangzhang tu' ('Map Which Measures One Zhang by Width and Length'). But the maps themselves, or copies, did last until at least the eighth century, when they were used by another great cartographer, Jia Dan, to compile an enormous world map. It took him seventeen years, and although this too has failed to survive, a reduced version of it was engraved onto stone in 1136 and can be seen in Shaanxi Provincial Museum. This is the map that caused Joseph Needham to proclaim: 'Anyone who compares this map with the contemporary productions of European religious cosmography . . . cannot but be amazed at the extent to which Chinese geography was at that time ahead of the West.'

Nearly 3,000 years after a deltaic scribe on the Euphrates impressed the world into clay, we have a revolutionary new geographical tool. Like so many game-changers, this one has been abbreviated into a neat acronym: GIS.

Geographic Information Systems connect data with geography. That's the simple bit. In the *International Encyclopaedia of Geography* of 2017, Kang-Tsung Chang defined GIS as 'a computer system for capturing, storing, querying, analyzing, and displaying geospatial data'. In Bo Huang's enormous, seventy-nine-chapter GIS tome of 2018, one of the contributors, Ming-Hsiang Tsou (a professor at San Diego State University's Department of Geography), described how GIS has come to cover three meanings: GISystems concentrate on the software and hardware for mapping and spatial analysis; GIServices deliver geospatial information, mapping services and spatial analysis via the internet and mobile devices; GIScience is 'question-driven' and uses scientific methods to understand geographical patterns, processes and relationships.

Data flows within GIS are facilitated by the geospatial cyberinfrastructure, a virtual world of geospatial computing and data coverage, wireless networks, geoprocessing services, geotagged data and time-honoured geographical knowledge. These interactions are a world away from the log tables, slide rules and paper maps of forty years ago. Geospatial data is geographical plankton: digi-food for Artificial Intelligence (AI) coded into devices that use machine learning – data-driven algorithms – to improve their tasking. In a geographical context, machine learning can automate prediction, clustering and classification in areas such as air pollution, delineating land-use types and identifying concentrations of social-media activity after a natural disaster. 'Deep learning', a subset of machine learning, makes it possible for software to train itself to perform tasks that include image recognition. Spatial patterns can be detected and images classified automatically.

Like a genie in the night, GIS has crept behind our screens. Apps on smartphones collect and edit field-work data; they geocode and upload text, photos and video that can be merged with map layers on the Web.

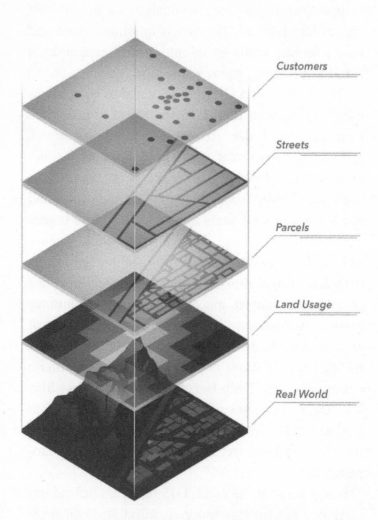

Customers

Streets

Parcels

Land Usage

Real World

*GIS can analyse spatial location and organise layers of
information into visualisations using maps and 3D scenes*

Touchscreens interact with virtual maps in the cloud. GIS business software can convert internet traffic into sales, create store-locator maps, reveal trading connections, coordinate distribution, run tracking systems, manage assets and liabilities. As we head for the third decade of the new millennium, GIS is so vast and versatile that there is scarcely a facet of modern existence that is untouched by it. From Chicago to São Paulo, Beijing to Chennai, GIS is driving millions of everyday decisions. Natural hazards, timber management, flood zones, wildlife habitats, resource management, land-use planning, transportation, health planning, military operations, farming and crime are all at the front end of a list of GIS applications. GIS has immense potential in a world of need: emergency relief to disaster zones can be delivered more efficiently and so can health care. GIS programs can be used to combat malaria and Ebola. One example is the work being undertaken by Andrew Tatem of the University of Southampton, who is investigating the use of 'de-identified' mobile-phone data to fill gaps in population data, a pressing need in countries where lack of statistics is making it much harder to tackle poverty.

Not for the first time, a tech acceleration is triggering alarms. It happened with the invention of steam power, with bicycles and then with the internal combustion engine. It happened with the arrival of farming (although few hunter-gatherer groups survived to tell their side of the story). AI will automate processes historically undertaken by human beings. And GIS has the potential to widen the geospatial tech gap. Geospatial cyberinfrastructure costs a lot of money and there are wide disparities between countryside and city – between LEDCs and MEDCs, between poor and rich. Vast swathes of Africa and Asia are not yet covered by Google Street View, while some cities are updated far more frequently than others. To benefit from GIS ervices, you need to be connected. Communities beyond the reach of geospatial cyberinfrastructure may not get access to services such as smart transport or humanitarian relief. In Ming-Hsiang Tsou's view, digital discrimination 'could trigger serious social problems and social unrest'. To counter this, the role of public-participation GIS (PPGIS) will become increasingly important.

Will GIS merely consolidate existing power relationships? Or can it be harnessed more positively to empower marginalised communities? The GIS project run by the Slovenian-born head of Spatial Collective, Primož Kovačič, was used to teach the citizens of two under-served settlements in Nairobi how to map their own communities. Kibera and Mathare house over 3 million people but were barely described on official maps. With GPS devices, Kovačič was able to 'turn people into data scientists'. The lack of water and power, health care and rubbish collection could all be mapped for the first time by 'citizen cartographers'.

Geospatial data and GIS are at the top of the twenty-first-century geographical toolkit. Every global issue needs them. Maps have always been the universal language of geography. We are all innate mapmakers. A few lines and dots in dirt or snow are enough to convey concepts of shared place and space.

More than thirty years ago, the great carto-historian J.B. Harley wrote of maps as 'mediators between an inner mental world and an outer physical world'. They are 'fundamental tools helping the human mind make sense of its universe at various scales'. In the new, interactive, 3D digital world of deep-learned GIS, we have the means to comprehend our futures.

SIX

Age of Geography

I n this final chapter, I'm going to celebrate the students and teachers whose collective geographical
knowledge is greater than any other sector of society.
Then I'm going to look at the menu of geographical
challenges facing humanity and ask how we move
forward. But first, the educators.

Geography teachers hold the world in their hands.
They are The Incredibles of school and university. Let
me tell you about Lucy Sprague Mitchell.

Seventy years before a team of geographers and
psychologists discovered that young children of all
cultures can work with map-like models, an American educationalist had come to a similar conclusion.
Mitchell was born in 1878 and educated at Radcliffe
College, a women's liberal arts college in Cambridge,
Massachusetts. It was an era of gender edginess: Radcliffe's educational prowess was a source of resentment

and anxiety at nearby Harvard College, where pro-
fessors feared co-education and the emasculation of
Harvard's 'purely virile' status. But this was the age of
Charlotte Perkins Gilman and Jane Addams, feminists
who showed the way for young women of imagina-
tion and energy. Mitchell moved from Cambridge to
California, where she became Berkeley's first dean of
women, but her true calling was in the east, in New
York. Three decades in Greenwich Village established
Mitchell as a teacher, a theorist on progressive school-
ing and founder of Bank Street College of Education.
In her remarkable book *Young Geographers: How
They Explore the World and How They Map the World*,
Mitchell insisted that 'even young children can and do
think in geographic terms'. She created a geography
curriculum for children aged from four through to
thirteen. Mitchell was an early advocate of school field
trips and mapmaking and thought of the neighbourhood around a school as an extension of the classroom.
She pleaded with her readers to 'think of geography
as a laboratory pursuit' and urged 'all teachers to hunt
with their children for source material, either in the
immediate world around them or wherever scientific
geographic data can be found, and to invent tools for
the study of the relationships inherent in these data'.

Led by Mitchell, geographical scholarship in the
USA might have taken a different turn, but in the

latter part of the twentieth century the academic discipline suffered decades of ill health, not least because Harvard dropped the teaching of geography. Harvard was followed by other leading American institutions. The cost has been high. A survey conducted in 1989 by the Gallup Organization revealed that 14 per cent of Americans could not find the USA on a map. One who took to the high ground during this episode of geo-illiteracy was a professor of geography at Georgetown University in Washington DC. Harm de Blij was born in the Netherlands, schooled in Europe, became an undergraduate in Africa and sat his graduate degrees in the USA. A brilliant communicator and academic, he wrote over thirty books and for forty years promoted geography through TV, journalism and lecturing. For de Blij, geography was the 'antidote to isolationism and provincialism'. In 1995 he was moved to warn that 'a general public not exposed to a good grounding in geography can be duped into believing all kinds of misinformation.' The post-truths of the twenty-first century are partly rooted in geographical ignorance and misuse.

In the USA, Canada and UK, geography is now taught from the first stage at primary school, when children are typically aged around six. In England, the National Curriculum at Key Stages 1 and 2 covers a range of subjects that include physical and human features of the landscape, geographical processes, map skills and GIS. In theory, no English seven-year-old should look up from a globe and exclaim: 'I never knew we had so many countries!' Or search an atlas for 'Nambia'.

Children in China and India are less well served. A paper published in 2015 by Xiaowei Xuan, Yushan Duan and Yue Sun revealed that in China geography was not taught at all in years one and two (that is, ages six and seven), and for years three to six the schools surveyed were desperately short of qualified teachers: 90 per cent of primary schools had no teacher with an educational background related to geography. Geography teaching in Chinese middle school is more systematic. A paper published in 2018 reported that most geography teachers had 'ample geography and interdisciplinary knowledge' and that most students were familiar with 'people, resources, environmental problems and climate change.'

In India, the lack of early geographical education is part of a wider learning crisis. India has the largest population of illiterate adults in the world, some 287 million. Around 40 per cent of children leave school before completing the eighth grade (at age fourteen) and without acquiring the basic skills of reading and writing. In secondary schools, most teachers have – in the words of an educational report – 'little background

Key to life on Earth: marine science

in geography and inadequate teaching competence'.

UNESCO lists a further twenty countries with the same kinds of learning crises. Many of the most needy countries are in regions such as sub-Saharan Africa, where exposure to environmental stress is already acute.

In a perfect world, every young child would have access to a basic geographical education. Given that children of school age have the ability to work with map-models, it makes sense to introduce them to the language of geography from their first year of formal education. Vital, interactive geographical concepts can accompany their journey into mathematics, language and play. Geography has much to do with playscapes, with the miniaturisation of reality and the creation of small-scale models. Labyrinthine cities, dendritic river systems, webs of transport and communication, colourful ecosystems and exotic landforms are well suited to the classroom and to local exploration. If the roots of geographical awareness are watered young, trees of knowledge will bear fruit. Educators have known this for a very long time.

Nurturing geographical imaginations at primary level rewards children with a world view which can be used as the foundation for secondary education, the springboard to adulthood and a role in decision-making that will determine the future of this troubled planet. Knowledge is its own multiplier: a little learning soon becomes a lot. A better understanding of our circumstances is the first step towards addressing the challenges. Many of our future problem-solvers can be found in universities. 'And', as the geoscientist Richard Alley pointed out in the preface to a new edition of his book *The Two-Mile Time Machine*, 'we have a lot of bright students. The knowledge of climate change can help motivate those students to move us toward a sustainable energy system.'

Geography has come a very long way since our distant forebears deployed their spatial skills to explore new environmental niches. It has evolved from a survival strategy to a science which informs multitudinous applications, from marine research to urban planning, housing, agriculture, security, business and much more. Human geographers are working to alleviate poverty, inequality and health crises; political geographers have been elected to head up governments; physical geographers share labs with climate scientists. Women and men with geography degrees are running multinational companies and global NGOs. Geographers are collecting data in tented Antarctic camps; they're implementing sustainable transport systems in cities; they're researching rural-to-urban migration in Africa and Asia and they're working with GIS to explore the Earth's surface in new and revealing ways.

Halley VI research station in Antarctica

Geography has always been in a state of reinvention; it is the river of knowledge, continuously modifying its own course. The shift in the 1950s from regarding geographical space as absolute to measuring space in relative terms of – for example – time and cost, opened a world of enquiry about human behaviour and space-adjusting activities.

Today, anyone with a primary or secondary school grasp of geography knows enough about the planet's major challenges to recognise that action is desperately urgent. The Earth system is in trouble.

The spines on my bookshelf are a chorus of mute shrieks: *Silent Spring, Six Degrees, Sea Change, Surviving the Century, The Last Generation, Heat, Hell and High Water, Seven Years to Save the Planet, Our Threatened Oceans, Field Notes from a Catastrophe, Our Final Century, Storms of My Grandchildren, The Uninhabitable Earth, On Fire* ... and those are just the tip of the (melting) iceberg. Back in 1997, the American geographer Robert Sack wrote in *Homo Geographicus* that 'we are now geographical leviathans. Our actions make things happen so extensively, quickly, and powerfully that we seem on the verge of rending the very fabric of nature, social relations and meaning.'

Over a century before Sack tapped his keyboard, the

pioneering conservationist George Perkins Marsh had warned in longhand:

> There are parts of Asia Minor, of Northern Africa, of Greece, and even of Alpine Europe, where the operation of causes set in action by man has brought the face of the earth to a desolation almost as complete as that of the moon ... and another era of equal human crime and human improvidence ... would reduce it to such a condition of impoverished productiveness, of shattered surface, of climatic excess, as to threaten depravation, barbarism, and perhaps even extinction of the species.

In a speech of 1847 to the Agricultural Society of Rutland County, Vermont, Marsh warned that 'climate itself has in many instances been gradually changed and ameliorated or deteriorated by human action'. And this was a decade or so before John Tyndall began to investigate links between the atmosphere's composition and climate change.

We cannot regain a 'natural' Earth. For centuries we have been disrupting the world's cycles of carbon, nitrogen and water. Ten thousand years ago, perhaps 0.1 per cent of mammal biomass was composed of people and domesticated animals. Today the figure has risen to about 90 per cent. The Anthropocene is with us. The first geological epoch to be defined by human intervention in the planet's natural systems marks the start of a new relationship between humanity and our spherical life raft.

On the brisk trek to civilisation and enlightenment we have increased food production, coped with pandemics and created national health systems. But we have also accumulated a rucksack of new difficulties. Depending upon how you unpack the contents of the rucksack, you can create different categories of priority. In 2015, the United Nations identified seventeen 'Sustainable Development Goals' and set the challenge that all should be achieved by 2030:

1 No poverty
2 Zero hunger
3 Good health and well-being
4 Quality education
5 Gender equality
6 Clean water and sanitation
7 Affordable and clean energy
8 Decent work and economic growth
9 Industry, innovation and infrastructure
10 Reduced inequalities
11 Sustainable cities and communities
12 Responsible consumption and production
13 Climate action

All of these goals are geographical. Leaving aside the considerable difficulties in tackling so many goals (they are broken down by the UN into 169 targets) in such a short period of time, and resolving potential contradictions (for example, the pursuit of higher global GDP alongside ecological imperatives), the list is nevertheless indicative of the scale of challenge we face in the coming decades. None of them can be addressed without a deeper, wider understanding of people, places and the environment. Some are less problematic to solve than others. All have implications for the future of humanity. Number 13 is the beast in the system.

The consensus on the cause of climate warming is solid. In the words of NASA: 'Multiple studies published in peer-reviewed scientific journals show that 97 per cent or more of actively publishing climate scientists agree: climate-warming trends over the past century are extremely likely due to human activities.'

Global warming is more difficult to bring to a halt than a car. Even if we ceased pumping carbon dioxide into the atmosphere today, the wretched stuff lingers. David Archer from the Department of Geophysical Sciences at the University of Chicago has estimated that – if we stopped emitting today – the atmosphere will still be polluted with 17–33 per cent of fossil-fuel carbon 1,000 years from now and that it will take 10,000 years to reduce it to 10–15 per cent. The carbon-dioxide tail is very long indeed: 7 per cent of fossil-fuel CO_2 will still be present 100,000 years from now.

In an effort to lessen the impacts of climate change, 195 countries adopted the first universal, legally binding, global climate deal at a conference in Paris in December 2015. The stated aim was to limit the increase in global average temperatures to 1.5 °C above pre-industrial levels. Unfortunately we have already seen a temperature increase of 1 °C. Even if the Paris cuts are implemented now, we may hit 1.5 °C as soon as 2030. There is a widely accepted view that at around a rise of 2 °C, many of our problems increase dramatically.

Policymakers have known this for decades. As long ago as 1972, Barbara Ward and René Dubos were warning in a report commissioned by the Secretary General of the United Nations Conference on the Human Environment that a rise in global average surface temperatures of 2 °C 'might set in motion the long-term warming-up of the planet'. The report was prepared with the assistance of a committee of 152 members drawn from fifty-eight countries. Dubbed on publication 'a Domesday Book of the kingdom of man', *Only One Earth* was a paperback success and a policy failure.

Fifty years have been lost. The atmosphere – one of the planet's four controlling components – is becoming less suitable for human life. As we saw at the beginning of this little book, the planet's four spheres – life, water, land and air – are so tightly meshed, so interactively entwined, that we cannot live on Earth without the complete quartet playing in harmony. There are immense pressures on the biosphere and the hydrosphere. But the climate issue has become so acute that attending to the atmosphere has shot to the top of the 'to do' list.

The UN's call in 2019 for 'unprecedented transitions in all aspects of society' was yet another warning. The transitions are urgent. The recent trends in emissions coupled to the failure of many countries to act with sufficient ambition on the Paris Agreement means that greenhouse gases are increasing at a rate consistent with hitting 2 °C by mid-century. Then 3 °C by 2070.

The implications of busting through the 1.5 °C Paris ceiling and reaching 2 °C were examined in an IPCC report of 2018. Heavy rains, hot days, droughts and floods will increase in intensity. By 2100, global mean sea level is projected to be 0.26 to 0.77 metres above 1986–2005 levels, *if* we restrain global warming to 1.5 °C. Reaching 2 °C adds an extra 0.1 metre and exposes an additional 10 million people to risk. At 2 °C, there is also an increased risk that the Greenland ice sheet and Antarctic marine ice sheets will become unstable, which could result in a long-term melt and multi-metre sea-level rise. Such is the interconnectedness of the world system that the rise from 1.5 °C to 2 °C would affect the entire biosphere, from wildlife to economic growth. The number of people exposed to climate-related risks and susceptible to poverty would, predict the IPCC, rise by 'up to several hundred million'.

The paradox of our age is that we're capable of taking extreme measures to protect ourselves in person yet appear indifferent to the survival of our species. It's probably time to acknowledge that the two are connected. We have denied and procrastinated. And now the economic system has collided with the natural system. Simultaneously, we have to prepare for the environmental changes already underway while working to avert an overheating planet. Both jobs carry an 'URGENT' sticker.

Among the environmental changes beginning to lap at our ankles are those brought by the water cycle. As we saw earlier, the water cycle is a closed system. But as temperatures rise and as ice melts, the volume of liquid water increases and becomes ever more dynamic. Some countries are more prepared than others. In England, where more than five million people are currently at risk from flooding and coastal erosion, the Environment Agency is preparing for resilience rather than protection. Strategies are being considered for a rise in global temperatures of 4 °C. Many communities may not be able to build themselves out of future climate risks and it may be better to help people relocate out of harm's way.

Across the North Sea in the Netherlands, the existing flood protection systems are probably the best in the world, but the latest estimates for sea-level rise along the Dutch coast are forcing upgrades to dykes and to storm-surge barriers, together with river-widening and coastal sand replenishment. Wherever possible, these climate adaptions are integrated with improvements to wildlife habitats and to public amenities: outside the town of Kampen on the River IJssel, a new water delta acts as a reserve for nature, with cycle tracks and trails, and the new flood channel is available for leisure boats. Nationally, a Climate-proof Together platform has been created and a central climate adaptation website, The Knowledge Portal, on which knowledge, tools and experience can be shared. Experts from Delta Programme 2019 (subtitle, 'adapting the Netherlands to climate change in time') are sharing knowledge with other low-lying countries such as Vietnam,

Bangladesh, Myanmar, the Philippines and Indonesia.

We have been here before, of course. Thousands of years ago, in eastern Asia, Yü the Great embanked and controlled the Yellow River so effectively that a vast, threatening floodplain became the fertile, populous kernel of China's first recorded dynasty. The endeavours of Yü – mythical or actual – hold the key to a resilient future. Yü's story shows that geoengineering can avert environmental disaster and in the process bring food, health and wealth to those previously at risk. We have a modern parallel in the Green New Deal, a plan that aims to build a more equitable future while cutting carbon emissions.

The Green New Deal is inclusive and sensible. In the process of protecting and regenerating the planet's natural systems, the economic model that has failed so many can be uprated to benefit all. The idea is to tackle the climate crisis alongside the creation of jobs and investment in health care, childcare, education and so on. One of the reasons that the UN's Sustainable Development Goals will miss their 2030 target is that the multilateral system has been undermined by an ailing world economy and disagreements over trade, currency movements and technology flows. At planetary scale, the UN sees a 'Global Green New Deal' as the best option for rebuilding the multilateralism necessary to tackle the receding 2030 SDG targets.

In the words of the UN's Richard Kozul-Wright:

> What is needed is a Global Green New Deal that combines environmental recovery, financial stability and economic justice through massive public investments in decarbonising our energy, transport and food systems while guaranteeing jobs for displaced workers and supporting low-carbon growth paths in developing countries – which is where the climate battle will be won or lost – through the transfer of appropriate technologies and sufficient financial resources.

We have to pause for a moment here to remind ourselves of the IPCC forecast that a rise in global temperatures from 1.5 °C to 2 °C will expose an additional 'several hundred million' people to poverty. The human cost of continuing with a business-as-usual model is truly unthinkable.

Kozel-Wright argues that the only way to deliver the public goods needed to achieve sustainable development for all 'is to create a well-funded, democratic and inclusive public realm at the global as well as the national level'.

There are, nevertheless, Green New Deals being hatched at national level, too. In the USA, the Green New Deal being backed by Democrats is a descendant of the emergency 'New Deal' programme Franklin D. Roosevelt deployed in the 1930s to propel the USA out of its Great Depression. Among its current enthusiasts are US Congress member Alexandria Ocasio-Cortez and author Naomi Klein, who sees America's Green New Deal as 'a moon shot approach to decarbonization, attempting to reach net-zero emissions in just one decade, in line with getting the entire world there by mid-century'.

At city scale, the Green New Deal has moved beyond a vision. At the time of writing, thirty cities have already peaked in emissions and are on track for net-zero. Another 100 cities have committed to climate action plans directed at meeting the 1.5 °C target. 'There is no other solution but a Global Green New Deal,' said the Mayor of Paris, Anne Hidalgo, in 2019. It is, she continued, 'the pivotal instrument to win this race against the clock. All decision makers must take responsibility in making it a reality.'

I'll finish where I began. Change is an indefinite deal. Land, air, water and life, the quartet of interactive components that allow us to drink from a mountain spring on a sunny morning, are geographical gifts not known to exist anywhere else in the cosmos. If we are to continue enjoying them, we'll have to know about the people, the places and the environments that make up our personal and shared worlds. Those worlds have always been in a state of change and we have always had to adapt. It is within our grasp to work for a good future. The UN's seventeen sustainable – geographical – development goals are out there as a global to-do list, with number thirteen, climate, asterisked for immediate action. The Green New Deals are out there as scaled templates for more generous, healthier futures. All of the sane options require populations, politicians, policymakers, business leaders, to grasp the basics of geography and to act immediately.

Never has geography been so important. On this finite orb, with its battered habitat, sustained in dark space by its intricate swirl of interconnected systems, we have reached a point in our collective journey where knowledge is the best guarantor of the future. Geography will keep us human.

Acknowledgements

A book like this arises from reading rather than writing, and I am – as ever – grateful beyond measure to the staff and resources of the Royal Geographical Society, the London Library and the British Library. I also thank the many Fellows and friends who shared their views during my time as President of the Royal Geographical Society. From undergraduate days, my good friend Martin Goodchild scrutinised an early draft. For his ever-methodical eye, I'd like to thank Dr Richard Crane for thoughts on this second edition. My literary agent, Jim Gill, provided peace of mind while I wrestled the world into a confined space. At Orion, my editors Alan Samson and Paul Murphy have been a constant source of expertise, encouragement and understanding. The efficiency of Ellie Freedman at Orion made it possible to revise the book and produce a second edition against a tight deadline. At home,

Annabel, Imogen, Kit and Connie have accompanied me through the peaks, passes and squalls of authorship. 'Thank you' will never be enough. While I was working on this book, a dear friend died. Douglas Rennie Whyte, husband, father, geologist, climber, cyclist, bee-keeper, topophiliac, was my companion on the geographical journeys and misadventures that set me on the long and winding road to this page. With Doug, I learned that trying for the impossible is more fun than settling for the probable.

Bibliography

Geography has attracted a cornucopia of scholarship and I've listed below a selection of sources I turned to while reading for this book. I've not included academic papers but have added a few key websites at the foot of the list.

General

Kish, G. (ed.), *A Source Book of Geography*, 1978

Richardson, D. (Editor in Chief), *International Encyclopedia of Geography, People, the Earth, Environment, and Technology*, 2017

Preface

Farrell. C., Green, A., Knights, S., Skeaping, W. (eds), *This Is Not a Drill: An Extinction Rebellion Handbook*, 2019

Thunberg, G., *No One Is Too Small to Make a Difference*, 2019

Wallace-Wells, D., *The Uninhabitable Earth: A Story of the Future*, 2019

1. *The View from L1*

Alley, R., *The Two-Mile Time Machine, Ice-cores, Abrupt Climate Change and Our Future*, 2000, 2014

Brooke, J., *Climate Change and the Course of Global History: A Rough Journey*, 2014

Castree, N., Demeritt, D., Liverman, D., Rhoads, B., *A Companion to Environmental Geography*, 2009, 2016

Fortey, R., *The Earth: An Intimate History*, 2004

Lenton, T., *Earth System Science: A Very Short Introduction*, 2016

Maslin, M., *Climate: A Very Short Introduction*, 2013

Matthews, J., Herbert, D., *Geography: A Very Short Introduction*, 2008

Poole, R., *Earthrise: How Man First Saw the Earth*, 2008

Roberts, N., *The Holocene: An Environmental History* (3rd edition), 2014

Woodward, J., *The Ice Age: A Very Short Introduction*, 2014

2. *Water World*

Digby, B. (series ed.), *Geography for Edexcel: A Level Year 1 and AS Level*, 2016

Digby, B. (series ed.), *Geography for Edexcel: A Level Year 2*, 2017

Dow, K., Downing, T., *The Atlas of Climate Change: Mapping the World's Greatest Challenge*, 2011

Gervais, B., *Living Physical Geography*, 2015

Macfarlane, R., *Landmarks*, 2015

Mack, J., *The Sea: A Cultural History*, 2011

Mayer, J. (ed.), *Alexis de Tocqueville, Voyages en Angleterre et en Irlande*, 1958

O'Grady, Patricia F., *Thales of Miletus: The Beginnings of Western Science and Philosophy*, 2002

Waddell, E., Naidu, V., Hau'ofa, E. (eds), *A New Oceania: Rediscovering Our Sea of Islands*, 1993

Wadhams, P., *A Farewell to Ice: A Report from the Arctic*, 2017

3. *Neuropolis*

Braudel, F., *A History of Civilizations*, 1987

Burdett, R., Sudjic, D. (eds), *Living in the Endless City: The Urban Age Project by the London School of Economics and Deutsche Bank's Alfred Herrhausen Society*, 2011

Chang, J., Halliday, J., *Mao: The Unknown Story*, 2006

Cresswell, T., *Place: A Short Introduction*, 2004

Dorling, D., Lee, C., *Geography*, 2016

Douglas, I., *Cities: An Environmental History*, 2013

Glaeser, E., *Triumph of the City*, 2011

Jacobs, J., *The Death and Life of Great American Cities*, 1961

Khanna, P., *Connectography: Mapping the Global Network Revolution*, 2016

Lahiri, J., *Unaccustomed Earth*, 2009

Latham, A., McCormack, D., McNamara, K., McNeill, D., *Key Concepts in Urban Geography*, 2009

Mehta, S., *Maximum City*, 2005

Tuan, Y., *Historical Geography of China*, 2008

West, G., *Scale: The Universal Laws of Life and Death in Organisms, Cities and Companies*, 2017

4. Yü, or How I Found My Inner Geographer

Baddeley, A., *Human Memory: Theory and Practice*, 1990

Blundell, G., *Nqabayo's Nomansland, San Rock Art and the Somatic Past*, Studies in Global Archaeology 2, 2004

Chuanjun, W., Nailiang, W., Chao, L., Songqiao, Z. (eds), *Geography in China*, 1984

George, A. (trans.), *The Epic of Gilgamesh*, 2003

Mulk, I., Bayliss-Smith, T., *Rock Art and Sámi Sacred Geography in Badjelánnda, Laponia, Sweden: Sailing Boats, Anthropomorphs and Reindeer*, Archaeology and Environment 22, 2006

Needham, J. (with Ling, W.), *Science and Civilisation in China*, Volume 3, *Mathematics and the Sciences of the Heavens and the Earth*, 1959

Sack, R., *Homo Geographicus: A Framework for Action, Awareness and Moral Concern*, 1977

Tilley, C., *A Phenomenology of Landscape, Places, Paths and Monuments*, 1994

Tuan, Y., *Topophilia: A Study of Environmental Perception, Attitudes and Values*, 1974

de Villers, G., 'From the Walls of Uruk: Reflections on Space in the Gilgamesh Epic', in Prinsloe, G., and Maier, C. (ed.), *Constructions of Space V: Place, Space and Identity in the Ancient Mediterranean World*, 2013

Waltham, C., *Shu Ching, Book of History: A Modernized Edition of the Translations of James Legge*, 1972

Wang, X., Jiao, F., Li, X., An, S., 'The Loess Plateau', in Zhang, L., and Schwärzel, K. (eds.), *Multifunctional Land-Use Systems for Managing the Nexus of Environmental Resources*, 2017

5. One to One

Barber, P. (ed.), *The Map Book*, 2005

Boas, F., *The Central Eskimo*, 1888

Brotton, J., *A History of the World in Twelve Maps*, 2012

Brotton, J., *Great Maps: The World's Masterpieces Explored and Explained*, 2014

Couprie, D., Hahn, R., Naddaf, G., *Anaximander in Context: New Studies in the Origins of Greek Philosophy*, 2003

Freeman, K., *Greek City States*, 1950

Harley, J., Woodward, D. (eds), *The History of Cartography*, Volume 1: *Cartography in Prehistoric, Ancient, and Medieval Europe and the Mediterranean*, 1987

Huang, B. (ed.), *Comprehensive Geographic Information Systems*, 2018

Imago Mundi, International Society for the History of Cartography, 1935

O'Grady, P., *Thales of Miletus: The Beginnings of Western Science and Philosophy*, 2002

Rovelli, C., *The First Scientist: Anaximander and His Legacy*, 2007

Schmidt-Glintzer, H., 'Mapping the Chinese World', in Mutschler, F., and Mittag, A., (eds), *Conceiving the Empire: China and Rome Compared*, 2008

Shore, A., 'Egyptian Cartography', in *The History of Cartography*, Volume 1, Part Two, Chapter 7, 1987

Wu, X., *Material Culture, Power, and Identity in Ancient China*, 2017

6. Age of Geography

de Blij, H., *Harm de Blij's Geography Book: A Leading Geographer's Fresh Look at Our Changing World*, 1995

de Blij, H., *Why Geography Matters More than Ever*, 2012

Klein, N., *On Fire: The Burning Case for a Green New Deal*, 2019

Lewis, S., Maslin, M., *The Human Planet: How We Created the Anthropocene*, 2018

Livingstone, D. *The Geographical Tradition*, 1992

Lowenthal, D. (ed.), *Man and Nature: Or Physical Geography as Modified by Human Action, George Perkins Marsh*, 1864, reprinted 1965

Morton, O., *The Planet Remade: How Geoengineering Could Change the World*, 2015

Sprague Mitchell, L., *Young Geographers: How They Explore the World and How They Map the World*, 1934, reprinted 1991

Ward, B., Dubos, R., *Only One Earth: The Care and Maintenance of a Small Planet*, 1972

Willy, T. (ed.), *Lending Primary Geography: The Essential Handbook for All Teachers*, 2019

Websites
http://www.antarcticglaciers.org
https://www.arctic.noaa.gov
https://www.bas.ac.uk
https://www.carbonbrief.org
https://www.cultureandclimatechange.co.uk
https://ec.europa.eu/info/
    energy-climate-change-environment_en
https://www.geography.org.uk

http://geographical.co.uk
http://www.ipcc.ch
https://www.istar.ac.uk
https://www.nasa.gov
http://nsidc.org
https://www.nationalgeographic.org/education
https://www.rgs.org
http://www.un.org/en
https://www.usgs.gov
https://www.worldbank.org
https://worldoceanreview.com/en

List of Illustrations

Index